girls in SCIENCE

National Science Teachers Association

Claire Reinburg, Director
Jennifer Horak, Managing Editor
Judy Cusick, Senior Editor
J. Andrew Cocke, Associate Editor
Betty Smith, Associate Editor

ART AND DESIGN
Will Thomas, Director
Cover Design - Toni Jones
Interior Design - Will Thomas

PRINTING AND PRODUCTION
Catherine Lorrain, Director

NATIONAL SCIENCE TEACHERS ASSOCIATION
Gerald F. Wheeler, Executive Director
David Beacom, Publisher

Copyright © 2008 by the National Science Teachers Association.
All rights reserved. Printed in the United States of America.
10 09 08   4 3 2 1

Cataloging-in-Publication data is available from the Library of Congress.

ISBN 978-1-93353-104-5

NSTA is committed to publishing quality materials that promote the best in inquiry-based science education. However, conditions of actual use may vary and the safety procedures and practices described in this book are intended to serve only as a guide. Additional precautionary measures may be required. NSTA and the author(s) do not warrant or represent that the procedure and practices in this book meet any safety code or standard or federal, state, or local regulations. NSTA and the author(s) disclaim any liability for personal injury or damage to property arising out of or relating to the use of this book including any recommendations, instructions, or materials contained therein.

PERMISSIONS
You may photocopy, print, or e-mail up to five copies of an NSTA book chapter for personal use only; this does not include display or promotional use. Elementary, middle, and high school teachers *only* may reproduce a single NSTA book chapter for classroom- or noncommercial, professional-development use only. For permission to photocopy or use material electronically from this NSTA Press book, please contact the Copyright Clearance Center (CCC) (*www.copyright.com*; 978-750-8400). Please access *www.nsta.org/permissions* for further information about NSTA's rights and permissions policies.

CONTENTS

About the Authors ix

Introduction xi

Section I
The Triad Story and Framework 1

Chapter 1
The Triad Story—A Science Education Community Navigating Gender Equity 3

Chapter 2
The Triad Framework—A Tool for Discussing Gender-Equitable Science Teaching 13

Section II
Exploring the Triad Framework Through Vignettes and Essays 27

Introduction 29

Chapter 3
Student Goals—Developing Girls Strong in Science 31

Contents 33

Introduction to the Chapter 35

Essay: Confidence to Explore 37
    After the Initial Eeewwww 39
    No Longer the Same 41
    Not Having Step-by-Step Instructions 44
    By the End of the School Year 46
    I Didn't Want to Produce the Same Fears 47

Essay: Familiarity With Tools 51
    Don't You Feel Powerful? 54
    The Real Microscope 56
    The Stopwatch as a Tool 57
    Safety Was a Concern 59
    I Don't Even Know How to Use a Saw 61
    The UV Bulb Can Be Changed by the User 63

Essay: Persistence Through Confusion 67
- Fun and Frustrating 70
- To Build and Rebuild 72
- Where to Draw the Line 74
- She Wanted to Do It Herself 76

Essay: Resilience to Failure 79
- I Shouldn't Have Come 82
- So That All the Bridges Fall 84
- Making Mistakes 86
- Watch Me! 88

Essay: Defending a Position 91
- I Assumed That Our Girls Would Feel Comfortable 93
- We Have Reason to Believe 95
- A Little Unnerving 97
- On a More Personal Level 99

Chapter 4
Science Goals—Envisioning Science in Classrooms 101

Contents 103

Introduction to the Chapter 105

Essay: Wonder About the Natural World 107
- Not What We Had Planned 109
- To Simply Marvel 111
- The Balloon Droops 113
- Above a Whisper 115
- Nothing to Do With the Club 117

Essay: Do Science to Learn Science 121
- Putting Sugar in Water 123
- I Learned How a Lava Lamp Works 125
- Nobody Knows What's Inside 127
- The Real Thing 129
- A Daunting Task 131

Essay: Think Critically, Logically, and Skeptically 135
Answers Are Not the Goals 138
Walking Encyclopedia 140
To Trust in Their Own Logic 142
There Is No Road Map 144

Essay: Use Evidence to Predict, Explain, and Model 147
The Most Difficult 150
Scientifically Dissatisfied 152
You Can Lead a Horse to Water 154

Essay: Build a Community of Scientists 157
The Strength of the Group 159
When Science First Was Really New 161
Have to Make It Right 163
To Help and Teach Each Other 165
Turning to Nadya 167

Chapter 5
Teaching Goals—Striving for Gender-Equitable Science Teaching 169

Contents 171

Introduction to the Chapter 173

Essay: Encourage Student Voices 177
I Watched in Awe 180
The Quieter Girls 182
I Have to Introduce Triad 184
See What Happens 187
Many People Got a Chance 188

Essay: Maintain High Expectations 191
The More We Expected 194
Talking in Questions 196
Can You Help Me? 198
Answering Student Questions With Questions 200
Theory Is Easy, Practice Is Difficult 202
Science Is Not a Priority for These Students 204
Accepting Stereotypes 206

Essay: Delegate Responsibility 209
- Keeping Your Hands in Your Pockets 212
- Does This Bridge Look Better Than It Did Last Time? 214
- I Could See How Much I Learned 216
- A Different Role 218
- No One Felt Uninvolved 220

Essay: Make Equity Explicit 223
- To Cunningly Mediate Equity 227
- Like Dad 229
- Talking About Equity 231
- Way Beyond Our Expectations 233
- Stop in My Tracks 235
- Anyone But the Boy 237
- My Own Tendency 239

Essay: Reflect to Improve Practice 243
- Back in the Classroom 246
- By Scoring When a Girl Participated 248
- Personal Development 250
- Resurrecting Socrates 252
- At First I Was Hesitant 254

Section III
Looking Forward and Learning More 257

Chapter 6
Taking Action 259

Appendix A: Facilitation Guidelines 267

Appendix B: A Few Words on Data Collection and Methodology 271

Appendix C: Literature Cited 273

Appendix D: Author Biographies 279

Acknowledgments 283

Index 287

ABOUT THE AUTHORS

Authors

Elizabeth (Liesl) S. Chatman
    Director of Professional Development
    Science Museum of Minnesota

Katherine Nielsen
    Co-Director, Science & Health Education Partnership (SEP)
    University of California at San Francisco

Erin J. Strauss
    Professional Development Project Lead
    Science Museum of Minnesota

Kimberly D. Tanner
    Assistant Professor of Biology
    San Francisco State University

All of these authors contributed equally and are listed in alphabetical order.

Contributors

J Myron Atkin
    Professor Emeritus of Education
    Stanford University

Marjorie Bullitt Bequette
    Professional Development Project Lead
    Science Museum of Minnesota

Michelle Phillips
    Evaluator
    Inverness Research

INTRODUCTION

This book is about promoting gender equity and, beyond that, equity on a larger scale within science education. It is for anyone who is engaged in science teaching: school teachers, professors, museum educators, school volunteers, and professional developers, to name but a few. The ideas, models, and voices contained herein come out of an extended professional development effort, known as Triad, that took place in San Francisco with funding from the National Science Foundation.

The Triad Framework for Equitable Science Teaching that emerged from the Triad community of practice serves as the architecture of this book. Encapsulating the influences, goals, and strategies that we found useful and further developed, the Framework is a conceptual tool that helped us analyze our work from three standpoints: students, teaching, and science. It has encouraged us to be mindful of many different levels of promoting equity within science teaching. The Framework began in graphic form, and its organization around the nodes of students, teaching, and science has enabled us to take those findings from relevant research on gender equity, teaching and learning, and science education that had the most profound impact on our thinking and to integrate and translate the findings into concise goals. These goals, in turn, were given life through the articulation of concrete, meaningful behaviors and teaching strategies. The Framework is the conscience of this book.

The heart of this book lies in the experiences of the people in the Triad community of practice. Our community included teachers, scientists, professional developers, students, and evaluators; each is given voice through the use of minicases or, as we refer to them throughout: *vignettes.* These vignettes are culled from first-person narratives, excerpts of transcripts from Triad's qualitative evaluation effort, and program records of events involving Triad students, teachers, and scientists. They are brief, often full of conflict and inner tension, and, at times, might feel to you, the reader, like passages from a story or fragments from an overheard conversation. Our hope is that the open-ended and ambiguous nature of the vignettes coupled with the accompanying questions for reflection will cajole and provoke you into new insights and greater depth.

This book supports professional development. Throughout the evolution of our work, we've defined professional development as the process of becoming the professional each of us wants to be. The foundation of our approach to professional development is the conviction that our beliefs as teachers are intimately tied to our actions—and vice versa—and are therefore central to any effort directed toward positive changes in teaching practice. It is through the analysis of our beliefs and our own practices that we as teachers give substance, depth, and consistency to the improvements we try to

INTRODUCTION

make in our work with students. Although it can be done individually, analysis and reflection of this kind often benefits from opportunities to consider the beliefs and actions of others in a collegial and supportive environment. As this book unfolds, the great extent to which we have interwoven practice, professional development, and evaluation will become self-evident.

This book is the story of one community. The Framework with its goals and strategies, the research cited, and the vignettes have all been used by our community and have proven themselves to be valuable. We hope they will be just as valuable to you.

A Book With Many Entry Points

Because we ourselves—even when we care deeply about a given subject manner—don't always have the time or the inclination to read an entire work of nonfiction from start to finish, we wanted this book to be usable in both a linear and a nonlinear fashion. You can pick it up and read it straight through, cover to cover, or plop it open to a page and dive in. This dive-in is particularly feasible in Chapters 3, 4, and 5. We are indebted to the authors of the *Birders Handbook* (Erlich et al. 1988) for inspiring this approach and to the creators of the Project 2061 *Atlas of Scientific Literacy* (AAAS 2001) for one of the most magnificent uses of purposeful graphic organization ever to grace the pages of science education literature.

Our text is written in three sections. Section I tells the story of the Triad community, its context, and the struggles that resulted in growth (Chapter 1) and then provides an overview of the Framework (Chapter 2). If you are linear, love stories, and like a lot of context, this is a great place to start.

Graphics and Iconography: Section II is organized around the Triad Framework for Equitable Science Teaching and its iconography. If you are a visual learner and approach books nonlinearly, this iconography will orient you to where you are and guide you to places of interest. From its earliest stages of development, the Framework was conceived and enacted in graphic form. Think Venn diagrams. The Venn diagram in Figure A is the first form of representation of the Framework and refers to students on the upper left, teaching on the upper right, and science on the bottom center. The origins of the iconography are described in Chapter 2. We've used this iconography and related graphics throughout this book because we all believe that thinking and understanding in many forms—text, symbols, illustration, and graphics—helps us to think and understand better. As Elliot Eisner puts it:

Introduction

The selection of a form of representation is a choice having profound consequences for our mental life, because choices about which forms of representation will be used are also choices about which aspects of the world will be experienced. ... Thus the paradox: A way of seeing is also, at the same time, a way of not seeing. (2004)

We've used the icon in Figure A to orient what you're reading to the Framework. We've also annotated links to other pages and chapters to help you find related topics of interest.

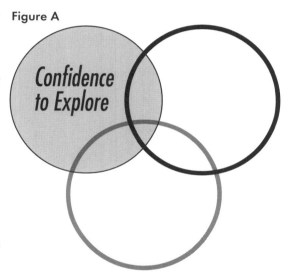

Figure A

Essays and Research: Research that had profound influences on our work is described in detail in the essays found in Chapters 3, 4, and 5. These three chapters are organized around the nodes of the Framework: Student Goals (Chapter 3), Science Goals (Chapter 4), and Teaching Goals (Chapter 5). As best we can, we've made efforts to introduce the reader to specific researchers and to point to their research articles—however, we want to make clear that our intent is not to present a meta-analysis or a comprehensive literature survey of research in all the fields related to gender, pedagogy, and science education. It is our intent to support our abilities to translate the ever-dynamic body of research into effective practices; our primary audience is the practitioner (and we include ourselves in that category).

Vignettes: The vignettes, described earlier in the introduction, are clustered around the goals of the Framework and accompanied by an essay also specific to a given goal. Our hope is that the vignettes will engage you in thinking about and discussing the dilemmas inherent in equitable science teaching. Each is accompanied by a set of questions for personal contemplation or communal discussion. The questions reflect the issues we grappled with as a community. We still wrestle with them. We invite you to use them in professional development contexts as well and have included more information about their use in Chapter 6 and Appendix A.

Our approach to writing this book has been to invite the reader to make the ideas presented herein your own and to remain true to the spirit of a dynamic and ever-evolving approach to teaching and learning. We're honored to have you join us in that process.

SECTION I
The Triad Story and Framework

CHAPTER 1
The Triad Story—A Science Education Community Navigating Gender Equity

Our efforts emerged from the backdrop of the science education reform movement in the 1990s. Then and now, a host of public policy documents and reports have elucidated the need for better education in the quantitative disciplines of science, technology, engineering, and mathematics (STEM). It becomes clearer and clearer that for economic health, the United States needs a larger and more diverse domestic STEM workforce. For everyday problem solving, scientific thinking is invaluable. For responsible citizenry in a world increasingly dominated by technology, scientific literacy on the part of the public is vital. These are some key rationales.

Yet, due to an array of factors, girls and women are still being steered away from STEM pursuits. In K–12, boys tend to be more confident about their math and science abilities than girls (Linn and Hyde 1989; Libarkin and Kurdziel 2003), and at an early age girls are more likely to develop negative attitudes toward science, resulting in self-doubt in their abilities (Steinke 1999). Girls often internalize disabling stereotypes, including the belief that computers, technology, and science are masculine and that there is a biological explanation for boys performing better in science, math, and technology (Gatta and Trigg 2001). In traditional K–12 science and math classroom environments, girls exhibit less self-confidence and are less assertive than boys (AAUW 1992; Fenema 2000). Boys tend to dominate discussions and interrupt girls and are more likely to be in physical possession of hands-on educational materials (Holden 1993). This environment is maintained by teachers' behaviors: Teachers ask higher-order questions of boys, are more likely to call on boys first, give boys more praise and substantive feedback, and, given the same task, will provide instructions for boys but show girls how to do it (Tobin and Garnett 1987). As a consequence, in K–12 STEM classes, boys are more likely than girls to have experiences with materials, experiences in problem solving, and opportunities to process these experiences through verbal communication, extensive acknowledgment, and feedback.

In high school, key issues expand to include course-taking patterns and lowered performance on national tests. For many African Americans, women, and members of other marginalized groups, stereotype threat around STEM raises anxiety and can affect negatively performance on exams, leading to avoidance of mathematics, sciences, and engineering (Adler 2007). Females continue to be underrepresented in advanced courses

SECTION I: The Triad Story and Framework

in both math and science during high school (AAUW 1998; NCES 2002), and females are more likely than males to stop taking mathematics courses after Algebra II. It makes sense, then, that males continue to receive mathematics and science honors and awards more frequently than females (Science Service 1998) and that females continue to score lower on the mathematics portion of the SAT college entrance exam (AAUW 1998). Similar gender differences are found in ACT and advanced placement mathematics and science test results (AAUW 1998). Moreover, these issues are compounded by race, ethnicity, class, and culture. Students of color are less likely to enroll in higher-level math courses, leading to an exacerbated and persistent gender achievement gap in K–12 mathematics within some racial and ethnic groups (Catsambis 2005).

At undergraduate, graduate, and postgraduate levels, fewer women than men enter college with an interest in pursuing science and engineering. In 2004, 26% of female first-year students intended to major in a science and engineering field versus 41% of males. By the time students reach graduate school, women earn only 34% of doctorates in science and engineering (Libarkin and Kurdziel 2003), and in 2004 women earned approximately 25% of computer science and information science undergraduate degrees and only 23% of the doctoral degrees (NSF 2007). These disparities manifest themselves in the workforce. In 2003, 28% of employed physical scientists and only 11% of employed engineers were women (NSF 2007). Women are also greatly underrepresented in information technology, systems analysis, software design, programming, and entrepreneurial positions (AAUW 2000). Some insist that a gender gap in science, technology, engineering, and mathematics no longer exists; the evidence indicates otherwise.

The Story of the Triad Community

Against this backdrop of inequity, the Triad effort began as a gender-equity initiative in San Francisco with funding from the National Science Foundation. Through Triad, teachers from the San Francisco Unified School District (SFUSD), scientists from the University of California at San Francisco (UCSF), and partnership specialists from the UCSF Science and Health Education Partnership (SEP) joined forces to understand and improve gender-equity dynamics in the local schools. Triad was so named because we sought to have a positive impact on three groups: scientists, teachers, and students.

Before we describe the Triad program, situating the effort within its organizational home, the UCSF SEP, will be helpful. Initiated in 1987 by UCSF Professor Bruce

CHAPTER 1: The Triad Story

Alberts, SEP is a longstanding partnership between UCSF and SFUSD. The specific mission of SEP is to promote partnership between scientists and educators in support of high-quality science education for K–12 students. To these ends, SEP develops and implements programs that support mutual teaching and learning among teachers, students, and scientists; promote an understanding of science as a creative discipline, a process, and a body of integrated concepts; contribute to a deeper understanding of partnership; and provide models and strategies for other institutions interested in fostering partnerships between scientific and education communities. The UCSF SEP has created several integrated models of partnership and more than a dozen programs that engage hundreds of scientists and teachers on an annual basis in 80% to 90% of the public schools in San Francisco. Triad, though, was the first SEP professional development project to address squarely issues of access and equity in science education.

In 1994, when Triad began, the program staff, teachers, and scientists came together because we shared a belief that science—its knowledge, culture, and habits of mind—belongs to everyone regardless of gender. Our initial approach was to bring together middle school girls, teachers, and scientists in a network of school-based girls science clubs. Our idea was to engage teachers and scientists in professional development, and they, in turn, would cosponsor science clubs and use these strategies in their club activities. We wrote a proposal around this concept, and in 1994 the UCSF SEP received an award through the NSF Experimental Program for Women and Girls (EPWG) to develop the Women's Triad Project, with Liesl Chatman serving as principal investigator. After-school Triad clubs were active in San Francisco's public middle schools and cotaught by teams of women teachers and scientists at the school site. Over the course of the next decade, the Triad effort expanded significantly in its approach, endeavoring to translate research into practice and, in the process, developed the Triad Framework for Equitable Science Teaching.

Initial Community Struggles

In the early years of our effort, we knew there wasn't equal access to science for all students and that gender featured in this inequity. We were committed to opening the doors to science, but changing the equity dynamics we'd grown up with was a big task. We read the research on gender and science, invited speakers from all over the country to come and share their expertise, and invested personal time and energy in engaging middle school girls in science learning through club activities. It didn't take long before we realized that meeting weekly with girls science clubs, making liquid

SECTION I: The Triad Story and Framework

nitrogen ice cream, and building bird houses just wasn't going to be enough. The problems that kept girls from being engaged and successful in science classrooms were too complex for a simple after-school science club to fix. As the Triad teachers and scientists worked with their science clubs and gathered with one another to learn more about gender equity issues in science education, they continued to say, "We understand the problems described in the research, and these problems are not going away. What can we do?" Clearly just knowing the research wasn't enough; more steps were necessary. We needed to move from having fun to being thoughtful. We needed to translate research findings in gender equity into explicit, concise goals and teaching strategies that our community could use. We needed to shift the focus from students to ourselves and our own teaching behaviors.

From Research to Practice

The Triad staff gathered in the fall of 1997 to discuss the development of specific, research-based goals. Because the program was about encouraging girls in science, we decided to begin with goals for girls. What did the research and our own experiences tell us were particular concerns? What would a girl who was strong in science look like? What kinds of intellectual and emotional resources would she have? She would be confident enough to explore on her own and with others. She would see confusion as an opportunity to learn and would have the persistence to work through problems. Failed experiments would provide information to this resilient girl; problem solving would be intriguing rather than defeating. She would use evidence to form and support her positions and would then have the courage not only to voice her opinion but also to address challenges to it and disagree with others. She would know how to use a variety of tools—both scientific and everyday—to explore and manipulate her world. Through this emerging vision, we distilled a set of Girl Goals (later referred to as *Student Goals*). In order to keep them accessible and memorable, we made them short and crisp. Girls would possess

- Confidence to Explore
- Familiarity With Tools
- Persistence Through Confusion
- Resilience to Failure
- Defending a Position with Evidence

The Triad community took these Girl Goals, which will be introduced in Chapter 2 and detailed in Chapter 3, and began to work with them in the context of science clubs and professional development workshops. We asked ourselves what kinds of experiences girls need in order to develop these traits, and with these discussions moved from activities that were fun and engaging to activities that were fun and engaging *and* purposeful in supporting the Girl Goals.

Clubs Begin to Evolve

With these new Girl Goals, things began to change. For example, science clubs had often done engineering activities, but now teachers and scientists were modifying those activities and experimenting with their own teaching in light of the goals. One such instance was modifying an activity centered around making bridges out of drinking straws and straight pins. In general, the initial challenge was, "Whose bridge can hold the most weight?" The lesson was structured around competition, but there was no particular attention paid toward instilling scientific or engineering habits of mind. Part of engineering is about testing designs in failure mode; it is important to know when and how things fail. The example of the Interstate 35W Bridge collapse in Minneapolis in 2007 tragically underscores this point. In Triad professional development and then in clubs, the bridge-design activity started changing so that the challenge became to test bridges in failure mode, to redesign them, and to increase the success rate of the bridges with respect to weight bearing across a given span. All girls had the opportunity to see their bridges fail, addressing the Girl Goal of resilience to failure. And all the girls had the experience of using this failure to make structural improvements in their bridges.

Real change, though, is not simple. It wasn't enough just to think about the girls and to modify the kinds of activities they did in the clubs. Girls spent the majority of their time in coed classrooms. The Triad science clubs needed to become environments in which the interactions between teachers and students and between scientists and students were just as critical as the activities undertaken. And our own interactions with students, again, often mirrored the inequitable behaviors described in the research literature. Concurrent with these realizations, many of the clubs went from being just for girls to being coeducational, spurred by California initiatives in the 1990s in affirmative action measures based on race, ethnicity, and gender. In the coeducational settings, problematic interaction patterns were even more evident. The teachers and scientists needed to change, but how? Our professional development needed to change, but how?

We began to regard our science clubs as equity teaching laboratories. Translating research into equitable practices in the context of coeducational instruction became

SECTION I: The Triad Story and Framework

the heart of our second proposal to the NSF, which was funded in 1998. The objectives of this broader effort, known as the Triad Alliance for Gender Equitable Teaching, were to develop a professional development program in gender equity and undertake a comprehensive research, documentation, and dissemination effort. This book is the fruit of that labor. The goals of the professional development program in gender equity were to

- engage teacher and scientist partners in iterative professional development focused on implementing equitable teaching strategies and engaging in reflective practice,
- develop a cadre of gender-equity leaders and mentors in the public school and the university, and
- enable teacher and scientist partners to pursue research into the effectiveness of their gender-equity efforts on girls.

Our team was joined by J Myron (Mike) Atkin, Professor of Science Education, Stanford University School of Education, who led our external evaluation group. Little did we know how profound the collaboration between our team of evaluators and professional developers would be.

We All Have to Change

One evening at a Triad professional development workshop in the first year of the second award, the adults were reflecting on their progress. The program staff was concerned about what we saw out in the clubs and began to question the efficacy of our own efforts. At the workshop, the team began to ask hard questions about how we all as adults were still inadvertently disempowering girls: "We've tried to provide the translation from research to practice you've been asking for. We've modeled pedagogical strategies. We've created concrete Girl Goals. But when we visit the science clubs, we still see patterns of interactions that foster inequities and limit access to rich science learning experiences for girls. Adults are taking materials out of girls' hands. Dominant girls are driving the group work. There seems to be more concern about girls having right answer than experiencing science in all its messiness. What's going on?" It was a tense moment for everyone, but the group members were committed to one another and to our work. As we talked, it became clear that we needed to discard the pernicious notion that someone was going to give us a magic bullet that would fix everything. In our Greek tragedy, there was no deus ex machina to come down and fix everything. "That's why we have a federal grant: because we are creating solutions that no one has come up with." We were

CHAPTER I: The Triad Story

the ones we were waiting for. That evening marked a turning point.

One club went back and videotaped a meeting and let students watch their own interactions. The club's scientist-teacher team then let the rest of us watch the video. We were all disturbed by what we saw. In one memorable scene, which involved making and testing small cars, the girls in one group sat silently with pieces of tape stuck on their fingers for the boys to use. The girls had become human tape dispensers. It was a rude awakening to us all. We had made progress through working with the Girl Goals, but it wasn't enough. It wasn't enough to talk about familiarity with tools. We needed to create an environment in which the tools stayed in the girls' hands.

Developing strong girls was not just about the girls; it was also about every one of us—the adults who worked with the girls—changing our own attitudes and teaching behaviors as well. Yet we had great difficulty combing the research to find out how to go about it. Yes, there was a wealth of resources, but to make the translation from research to practice was a monumental task. Strategies were spread far and wide, embedded in articles, written for the researcher and not the practitioner, and often described methods that were sufficiently complex to be impractical for an inner-city class of 40. Real change was up to us. If the girls in Triad were going to develop in the ways outlined in the Girl Goals, we adults needed some Teaching Goals of our own. As adults, we had to believe and expect that every one of the girls could become confident, persistent, and resilient. This meant we would have to let them explore, struggle with confusion, and even fail at some tasks. We had to trust the girls to ask important questions and do meaningful work; we had to let the girls be leaders. And, finally, we had to address our own fears and discomfort about talking explicitly about equity. We had to talk with one another about our own behaviors as adults; we had to talk with the girls about group dynamics in the clubs; and we had to take some club time to talk about society at large and what the girls saw and experienced in their schools, neighborhoods, and homes. Again, we distilled a set of goals for ourselves, a set of Teaching Goals, which will be introduced in Chapter 2 and detailed in Chapter 5:

- Encourage Student Voices
- Maintain High Expectations
- Delegate Responsibility
- Make Equity Explicit
- Reflect to Improve Practice

Again we poured over the research. And again the clubs changed. Our own actions changed as well. As a community, we were taking responsibility for figuring out where we needed to go next and how to get there. The goals quickly translated into teaching strategies that, in turn, became rallying cries. "Keep your hands in your pockets!" "Answer questions with questions!"

But What About the Science?

Fortunately, Triad had keen and thoughtful observers—our evaluators from Stanford University, Mike Atkin and his team of doctoral students, who had joined us with the new grant. The evaluation team members were an integral part of our community and very supportive of the work Triad was doing. They spent many hours watching and recording what happened in adult meetings and in clubs and thinking deeply about implications. At our end-of-the-school-year evaluation retreat in 1999, they came to the staff with a critical observation. They noted that it was great that Triad had Girl Goals and Teaching Goals, but this project was about girls in science. The evaluation team was observing lots of growth on the part of all the participants, but the science itself still seemed to be about fun. Now fun is grand—but shouldn't the girls, asked the evaluation team, be developing scientifically as well? In a fashion similar to those that had preceded it, the pendulum swung, and we began the process of creating Science Goals.

Although it may seem as though the Science Goals would have been the first and the easiest to develop, that was not the case. Partly, this was an issue of professional cultures and partnership. Triad brought classroom teachers who were responsible for teaching a wide sweep of science ideas together with scientists who were specialists in areas like neuroscience or cell biology or biomedical engineering. The cultures of their professional worlds were quite different, and agreeing on what was most important for girls to know about science was not an easy task. (For a detailed perspective on these professional cultural issues, see Tanner et al. 2003.)

By this time, the entire Triad community was working to refine the goals and associated strategies. Veteran participants had joined the staff in designing and conducting professional development. We re-read national standards documents and poured over the hot-off-the-press *Inquiry and the National Science Education Standards* (NRC 2000), which several of our good colleagues were involved in. By engaging in joyous and thunderous debates and trying on many versions of language, the community finally agreed on a set of goals—and then completely revamped them the following year! We came to consensus that what was common to all science were ways of thinking and doing that resulted in new understanding grounded in evidence. Clubs

CHAPTER I: The Triad Story

might explore many different kinds of science topics, but in every club girls should be thinking and acting scientifically. They needed to work in community to build their knowledge and skills; they needed to do experiments rather than just watch or hear about them; they needed to develop scientific habits of mind; they needed to experience the natural world as a source of wonder; and they needed to learn how to use evidence to make predictions, explain phenomena, and develop models. In the end, we distilled a set of Science Goals, introduced in Chapter 2 and detailed in Chapter 4, that had great traction:

- Wonder About the Natural World
- Do Science to Learn Science
- Think Critically, Logically, and Skeptically
- Use Evidence to Predict, Explain, and Model
- Build a Community of Scientists

With this final goal set, the Triad Framework emerged.

CHAPTER 2
The Triad Framework—A Tool for Discussing Gender-Equitable Science Teaching

In chapter 1, we described Triad and the Framework from a community perspective. In this chapter, we'll describe how the Framework evolved from a more theoretical standpoint. We'll introduce its anatomy and initiate a more detailed discussion of its goals and the research that informed them. Right up front, we need to be clear that the Framework is intended to be dynamic and not definitive—a point we will return to at the end of the chapter.

Evolution of the Framework

The Graphic Story

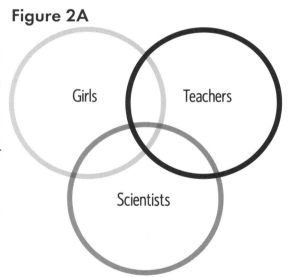

Figure 2A

As Triad came into being in 1993–1994, we wanted a name that would let people know that our community had three types of people—teachers, scientists, and kids—and that we had learning goals for each. So the name *Triad* was coined, and the initial graphic representation of the Venn diagram of three circles (Figure 2A) soon followed. This schematic, though, first came to us through a chalkboard talk at Berkeley's Lawrence Hall of Science from Larry Lowrey, professor emeritus of the University of California at Berkeley and the principal investigator of the Full Option Science System (FOSS) in the early 1990s. He used this Venn diagram to describe the knowledge arenas needed for effective science teaching: knowledge of students and learning, knowledge of effective teaching methods and pedagogy, and knowledge of the discipline of science. Its elegance has proven extraordinarily helpful, and you will see us use it as an über-organizing feature of this book. Within the Triad program, we consistently presented the icon in color, using the primary colors of light: green, blue, and red. We did this because, when all of the colors in light are combined, there is total illumination. It is a metaphor for our aspirations.

13

SECTION I: The Triad Story and Framework

Triad's Theory of Action and the Goal Sets

As our story in Chapter 1 evidences, as a community we struggled with how to translate research into tangible actions that educator scientists could practice with students. Knowing the research helped us to be aware of the problem of gender inequity, but it didn't automatically help us to see interaction patterns or to help us change our own behaviors. Our theory of action was that engaging teacher and scientist partners in a professional development cycle of preparation-action-reflection, based on clear equity goals, would enable us to increasingly implement equitable teaching strategies and reflect on our practice. First articulated in our renewal proposal to NSF in 1998, this theory of action was greatly influenced at the time by the work of Susan Loucks-Horsley and colleagues in their book *Designing Professional Development for Teachers of Science and Mathematics* (Loucks-Horsley et al. 2003).

Goals were fundamental to this process. Did we prepare lessons based on equity goals? Did we put them into action? What did we find when we reflected back on the extent to which we met our goals? We found that when we distilled our goals into easily remembered nuggets around the nodes of the Framework—student goals, teaching goals, and science goals—we got more traction in positively changing our actions and behaviors and the capacity to see and reflect. Thus, for each of the three nodes, we articulated a goal set. Each goal set includes four or five brief phrases that helped us to maintain a tight focus on behaviors and attitudes—such as the Student Goal, Persistence Through Confusion—that our work revealed were critical.

A Word About Grain Size

As we worked through creating the framework, levels of complexity and detail emerged at each step as we established clearer and clearer targets of what gender-equitable teaching and learning might look like. We came to think of these layers as grain-size issues. We used three basic "grain" sizes, analogous to rocks, pebbles, and sand. The largest grain size, that of rocks, is the goal set—the Student Goals, Teaching Goals, and Science Goals seen in Figure 2B—in the three major areas of students, teaching, and science. The goal set helps us to tease out questions such as "Are we talking about behaviors and attitudes linked to students and learning, to teachers and pedagogy, or to scientists and the scientific endeavor?" Within each goal set, the next-smaller "grain" size, that of pebbles, is the individual goal within the goal set. At this level, we ask, "What is the compelling argument that this is a gender-related issue?" Thus, we are concerned with relevant research and the rationale as to why we have focused on a particular goal. Finally, the smallest here-are-some-specific-actions grain size, that of sand, is at level of the strategies related to

CHAPTER 2: The Triad Framework

a given goal. These are the pragmatic actions that we can practice, the habits of mind that we can instill in ourselves and others, that turn research into practice.

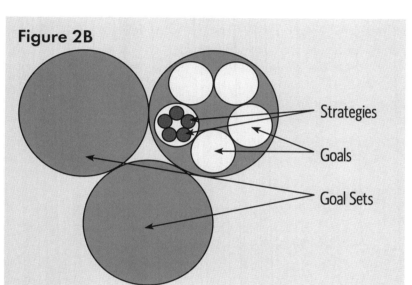

Figure 2B

This chapter deals with goal sets. Each of the three chapters in Section II addresses a goal set, and thus the discussions in these chapters will be at the level of the individual goals within the goal set and the accompanying exemplars/strategies. Section II is what we really consider to be the heart of the book; it's where we'll go into great depth, from theory to methods to implementation. These chapters will also include a collection of true vignettes that illuminate the complexities, challenges, and dilemmas of putting the ideas of the Framework into practice. These vignettes and where they came from are described in the introduction to Section II. Before we go there, it will be informative to return to the goal sets.

The Anatomy of the Framework and Key Research Influences

The Goal Sets

As we've indicated, for each of the goal sets mentioned above—students, teaching, science—the Framework assembles a group of concise goals to consider in promoting gender equity in science education. These goals were drawn from and synthesized by the Triad community from research, standards documents, effective-practices literature, and our own experience and data analysis. We wanted to find what specifically was key to changing the gender-equity climate in science classrooms. What were the problems we needed to address? The behaviors we needed to change? The attitudes we wanted to instill? The vision we intended to create?

SECTION I: The Triad Story and Framework

Student Goals

As we began looking at the research in gender equity in the mid-1990s, over and over we saw descriptions of interaction patterns between teachers and students and between students and students that were mediated by gender. We saw that these interaction patterns restricted opportunities for girls that were key to science.

One of the fundamental aspects of the practice of science is its orientation around process. Our students need concrete, provocative experiences with materials in situations where they can express curiosity, explore natural phenomena, perform experiments, solve problems, and communicate observations, predictions, and conclusions.

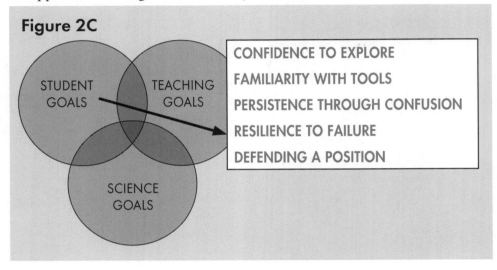

As they do so, they must persist through confusion, ask questions, and defend positions as they talk about science with their peers. As we reviewed research, we found ample evidence that the prevailing culture of the middle school science classroom greatly restricts these experiences for girls. In a mixed-sex environment, boys express more self-confidence and assertiveness than do girls (AAUW 1992), are more likely to dominate discussions, to interrupt girls, and to be in physical possession of educational materials (Holden 1993). Hence, boys are more likely than girls to have concrete experiences in science and greater opportunities to problem solve, thereby fostering self-confidence and promoting a richer understanding of abstract concepts. Since self-confidence for all students is most closely linked to achievement in math and science (Linn and Hyde 1989), it is unsurprising, then, that middle school is the point at which girls' confidence in science declines, followed by a decrease in interest and achievement (AAUW 1992). Our local data bore this out; statistical analysis of 6th- through 11th-grade scores on standardized math achievement tests (the 1996 Math Concepts and Applications portion of the Comprehensive Test of Basic Skills) showed that girls in the San Francisco Unified School District were not achieving at the same levels as their male peers.

CHAPTER 2: The Triad Framework

Although a loss in confidence during the middle school years is a critical factor leading to reduced enrollment of girls in high school science electives, the problem is not limited to the adolescent period. As they continue their education through high school and into college, young women encounter the same cycle of discouragement. Even the female survivors who go on to graduate school and beyond frequently suffer with inappropriately low self-confidence and/or a low sense of efficacy. The attrition rate of females versus males in one UCSF research department evidences the disparity. Of the males leaving the department, 83% exited with a PhD while only 47% of the women did so. At the other end of the exit profile, only 10% of the men versus 26% of the women left with no graduate degree. For those who persevere at the graduate and postgraduate level, women scientists report setting less ambitious goals than their male counterparts, and they are also less likely to be in positions of leadership in the scientific community (Vetter 1992). The ultimate effect is that females are strained through an increasingly fine mesh that separates the discouraged from the few who still have their confidence intact.

Thus, the student goals (see Figure 2C) are rooted in the predispositions and metacognitive skills needed in science. However, they go beyond standards documents and are informed by research in education, sociology, and gender studies. They are intended to promote the internalization of attitudes and actions, such as resilience and effort, that are vital to science but, research indicates, aren't fostered in girls. They point out the ways that individuals should engage in science but, because of gender-based interactions, don't.

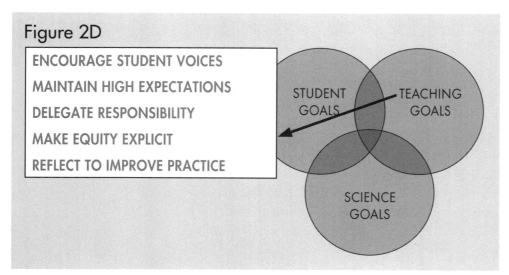

Figure 2D

- ENCOURAGE STUDENT VOICES
- MAINTAIN HIGH EXPECTATIONS
- DELEGATE RESPONSIBILITY
- MAKE EQUITY EXPLICIT
- REFLECT TO IMPROVE PRACTICE

Teaching Goals

The environment we described above is maintained in a circular fashion by our teaching behaviors: We ask higher-order questions of boys, are more likely to call on boys first, give boys more praise and substantive feedback, and given the same task, provide instructions for boys but show girls how to do it (Tobin and Garnett 1987). Even though girls are less likely to interrupt, if they do, we're more apt to reprimand them. We tend to give girls less critical feedback—at worst we give them irrelevant feedback like "Neatly done!" accompanied by a smiley face regardless of the quality of their work, good or bad. Girls are also likely to talk at a less-abstract level in coeducational groups (Martinez 1992). In our efforts to be sure students do it "right," we hover and rescue (Cohen 1994; Gordon 1995), usurping their sense of authority and efficacy. We ask a simple question, immediately call on the first hand raised (usually a boy), and then lament that other students aren't participating. It takes conscious and focused effort for us to change these patterns. Those of us who are postsecondary science educators usually have no training in educational strategies; we often unwittingly repeat the same disparate treatment of males and females that students experience in precollege science education.

The teaching goals (see Figure 2D) help us to create a vision of the professionals we want to be. They are about facilitating student effort and apprenticeship, promoting and maintaining academic rigor regardless of gender, and being explicit about equity. Influences in this area included Mary Budd Rowe's work on wait-time (1974), Paul Black's emphasis on thinking about the questions we ask students (Black and Wiliam 1998; Black et al. 2004), Elizabeth Cohen's explorations into delegating authority (1994), and, more recently, Lauren Resnick's synthesis of a variety of research studies in "From Aptitude to Effort: A New Foundation for Our Schools" (1995).

Science Goals

The kind of in-depth inquiry called for in the National Science Education Standards (NRC 1996, 2000) and by Project 2061 (AAAS 1993, 2001, 2007) asks that learners engage in the habits of mind and authentic practices in which scientists engage. These documents make connections between the kinds of questions that scientists ask and the forms of investigations they use to address their questions and develop explanations. They emphasize data gathering, evidence, logic, analysis, and communication of results.

The Science Goals (see Figure 2E) are intended to encapsulate the discipline-based habits of mind used in science and are true to the intellectual intent and culture of science. They reflect the practices engaged in by practitioners both with respect to community structures and the authentic activities of scientists. Coupled with the Student Goals, they relate strongly with the essential features of classroom inquiry (NRC 2000):

CHAPTER 2: The Triad Framework

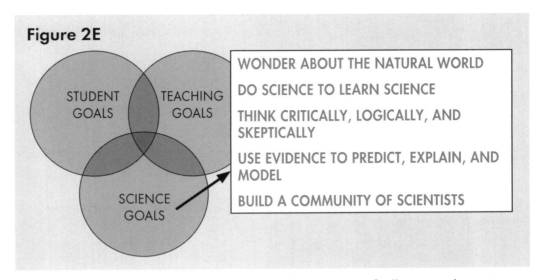

Figure 2E

- The learner is engaged—mentally engaged—in a scientifically oriented question.
- The learner gives priority to evidence in responding to a question.
- The learner uses evidence to develop an explanation.
- The learner connects explanation to scientific knowledge.
- The learner communicates and justifies explanation.

Integrated Nature of the Goal Sets

Ultimately and collectively, the goals are about engaging all learners in science. The goals are formulated to be mutually supportive when integrated. For example, you may focus on the Student Goal of Defending a Position and the Science Goal of Use Evidence to Predict, Explain, and Model, and the Teaching Goal of Encourage Student Voices. These goals imply three things: one, that girls actually have evidence and know what constitutes evidence; two, that they take a position based on that evidence; and, three, that they have an opportunity to share their evidence and voice a position. You may then turn again to the Science Goals and see that Do Science to Learn Science sets the stage for experimentation and see that Use Evidence to Explain, Model, and Predict involves analyzing experimental data that will help students to formulate a position. Teaching Goals that will create opportunities to foster these behaviors are Delegate Responsibility and Encourage Student Voices. Delegating responsibility can be done by returning to the Science Goals and allowing groups of students to Build a Community of Scientists, designing experiments around a central topic or phenomena that collectively will produce a variety of results. Creating a forum for presenting their results and interpretation not only encourages student voices, it fosters Think

SECTION I: The Triad Story and Framework

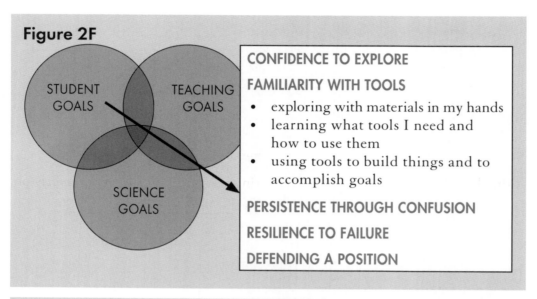

Figure 2F

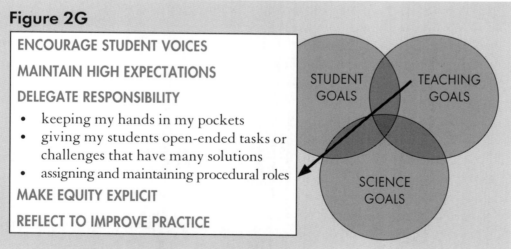

Figure 2G

Critically, Logically, and Skeptically as well. In this manner, the Framework is mutually reinforcing and can inform instructional decisions and practice.

From Goals to Actions: Strategy Exemplars

Finally, there is a last layer of graphic representation that looks like a flow chart. In Figure 2B, these are the smallest circles; in Figures 2F and 2G, they are the bullet points that relate to a specific goal. This layer of strategies creates a vision of ways

CHAPTER 2: The Triad Framework

of being, thinking, and acting in a science education environment. It is the heart of the translation from abstract to concrete, from theory to practice. For each goal, we articulated what we found to be a critical strategy, habit of mind, or behavior pattern that put flesh on the bones of a given goal. For example, it's fairly easy to say, "I have high expectations of my students." And we earnestly believe that we do. And yet, when we look at our actions, they betray us. We might give in when a student backs away from the equipment, expecting that we will rescue her by doing it for her. Our action, that of rescuing, conveys the expectation that she can't do it.

A simple teaching strategy related to delegating responsibility, then, is to merely put one's hands in one's pockets as we talk to students who have questions about equipment. In this case, we delegate responsibility for doing the work to the student, maintaining the expectation that she can do it.

A Dynamic Translation Tool, Not a Definitive Collection

It is critical to understand from the outset that the goals and their related strategies are not intended to be definitive or exhaustive. They are—and thus the Framework is—a practical starting point, a tool to translate theory into practice, a way of moving from the abstract to the concrete. The Framework is an invitation to explore and purposefully change the interactions and behaviors in the science-learning environment to promote engagement of all individuals within that environment.

The original graphic used to describe our community evolved in Triad to have many functions. We first used it to describe our participants (teachers, students, and scientists). In time, our colleagues from Stanford who served as our evaluators used it as an organizing structure to assess our community's progress. Ultimately, we used it to frame our goals. It's important to remember that we took small steps to get there and that we didn't have a clear vision in 1993 of where we were going. We only knew what our theory of action was, kept our eyes on solving the problem at hand, and acknowledged that no one had the magic bullet. We accepted that we were the ones we were waiting for.

The Triad structure has been harnessed by those of us who have used it to explore other content disciplines and other arenas of equity such as English language learning. And in the course of writing this book, we've refined the strategies, revisited research, reframed our discussions to address race, class, and culture, and through it all engaged in lively and healthy debates with each other.

SECTION I: The Triad Story and Framework

An Invitation

When we saw Pinky Nelson introduce the *Atlas of Scientific Literacy* (AAAS 2001) at the 2002 National Science Teachers Association's national convention in Saint Louis, he invited the audience to sit down with the Atlas accompanied by a nice pen, a ruler, some Wite-Out, and good bottle of chardonnay. It was an invitation to make one's own sense of the Atlas and a nod to the notion that there are many ways that concepts can be organized. We hope that you will do the same with Triad Framework—that you will morph these graphic tools into your own purposes and make them your own.

CHAPTER 2: The Triad Framework

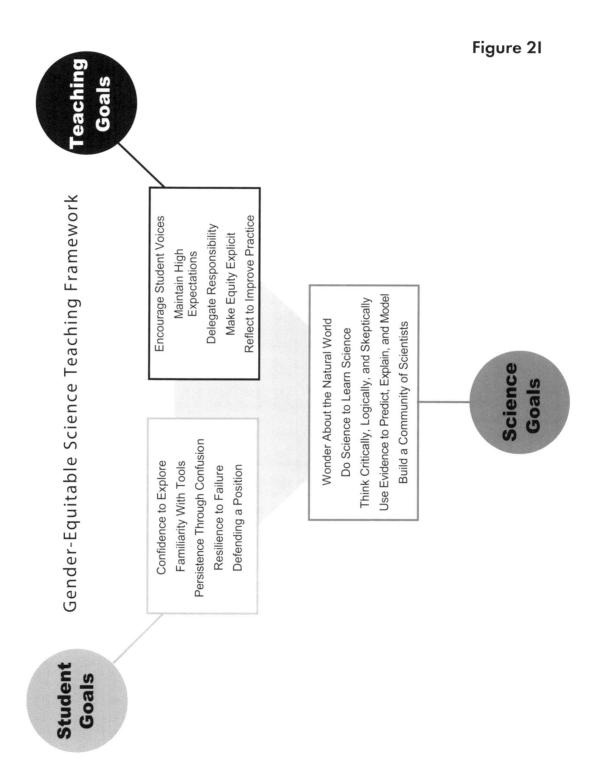

Figure 21

Girls in Science

23

SECTION I: The Triad Story and Framework

Figure 2J

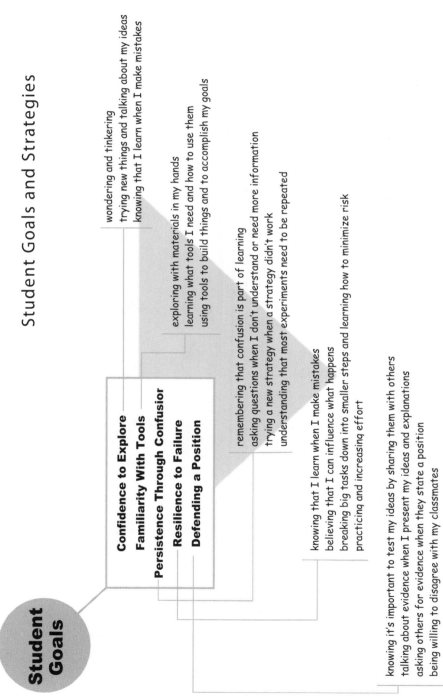

CHAPTER 2: The Triad Framework

Figure 2K

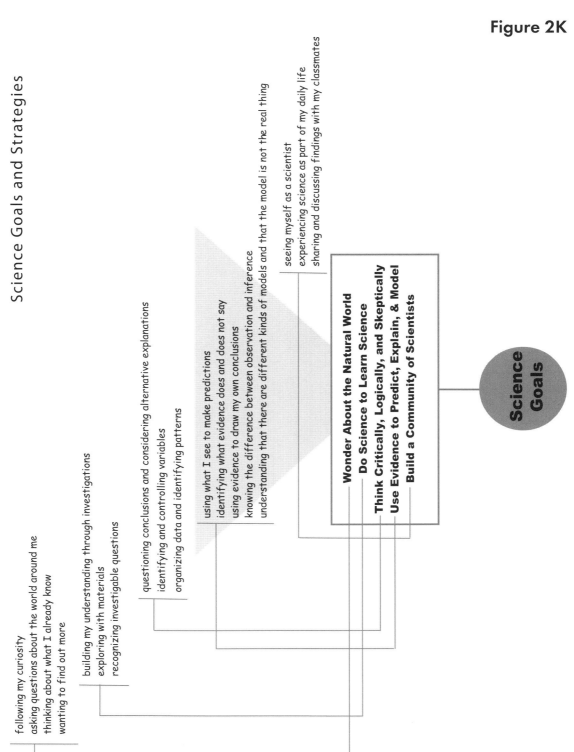

SECTION I: The Triad Story and Framework

Figure 2L

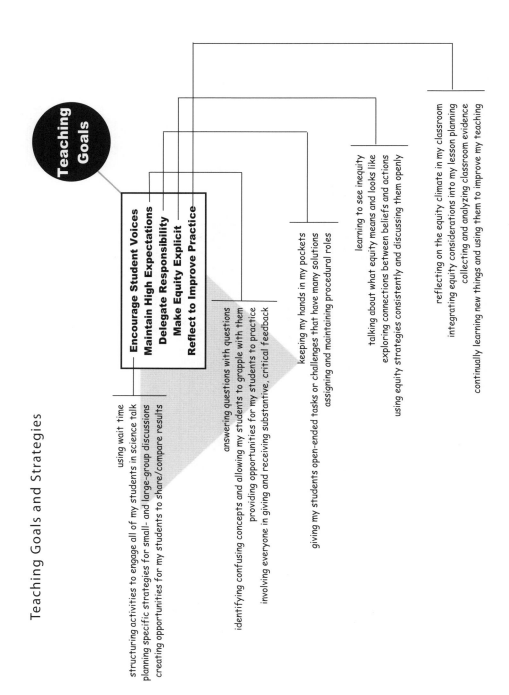

SECTION II:
Exploring the Triad Framework Through Vignettes and Essays

INTRODUCTION

The Structure of the "Heart" Chapters
The purpose of this introduction is to prepare you to move quickly into the heart of this book: the voices of the Triad community and the depth of what we learned collectively about gender equity in science education. To illuminate these two pieces—voice and depth—we've used two elements: vignettes and essays. If you are using this book as an individual, this short chapter will serve as the platform for you to dive right in. If you're using this book in the context of professional development programs, more detailed information about how to use the vignettes and essays is in Chapter 6 and Appendix A.

Rhythm: Essays and Vignettes
Each of the following chapters—chapters 3, 4 and 5—is organized around one of the goal sets of the Triad Framework for Equitable Science Teaching, as described in the previous chapters. The chapters are like musical measures. Each chapter includes a detailed overview of the goal set followed by a set of clusters around each goal. Each cluster contains two things: first, an essay on the goal, in which the goal is unpacked and connections are made to relevant research and policy that contributed to the formulation of the goal, and second, a set of brief first-person minicases—vignettes—that are related to the given goal. Each vignette is accompanied by set of reflection questions and links to related vignettes throughout the book. Lastly, we've used the Triad iconography described in Chapter 2 much as a field guides uses symbols to aid readers.

The vignettes are taken from a variety of written artifacts produced throughout the formal Triad program (see Appendix B for more information). In most if not all instances, they are vignettes from longer material produced by participants, staff, and evaluators. These vignettes have been pulled from first-person reports and reflections written at workshops and retreats, from transcripts associated with the qualitative evaluation conducted by the Stanford Evaluation Team, and from program records of events involving Triad students, teachers, and scientists.

The vignettes are intentionally brief and provocative, leaving many details vague or omitted. They are not developed cases but rather quick entry points into thinking or talking about tough, challenging topics because their ambiguity demands it. To foster contemplation, each of the vignettes is accompanied by several open-ended questions to prompt reflection and discussion. By grouping the vignettes into clusters, articulating reflection questions, presenting relevant information close by through essays, and including links to other related material, our hope is that we have made the material accessible to a broad variety of contexts and applications—and that it becomes personal.

SECTION II: Exploring the Triad Framework

Connections: Iconography and Links

The goals and strategies presented throughout this book are mutually reinforcing and highly interdependent. This is what gives them their power and coherency. In order to support an understanding based on this interdependency, we've used consistent structural devices and visual patterns to help you to snoop around out of sequence, to return easily to material you've already read, and to forge connections.

Figure A shows the general iconography for the Triad Framework as described in Chapter 2. This Venn diagram, in turn, is used throughout chapters 3, 4, and 5 with one difference: The specific goal appears in the corresponding goal circle. For example, the goal arena for Student Goals is always shown in the upper left of the Triad Venn diagram. One of the five Student Goals is Confidence to Explore. The icon shown in Figure B has the goal written in the upper-left sphere and will accompany any essay or vignette that is primarily related to the goal, Confidence to Explore.

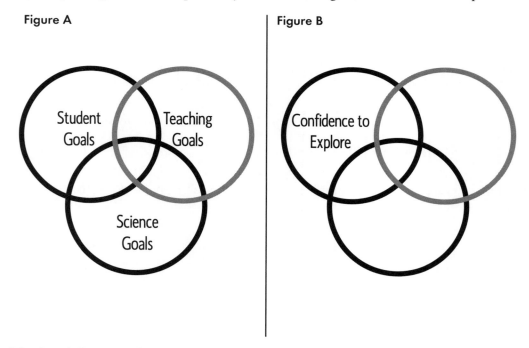

The "Links" section follows the set of reflection questions for each vignette and refers you to other vignettes whose issues intersect with the one you've just read. In most cases, the links will take you to a different node of the framework and do so through the title of the related new vignette. This new vignette, in turn, will be nested within a goal with its own relevant essay.

CHAPTER 3:

Student Goals—Developing Girls Strong in Science

SECTION II: Exploring the Triad Framework

Figure 3A

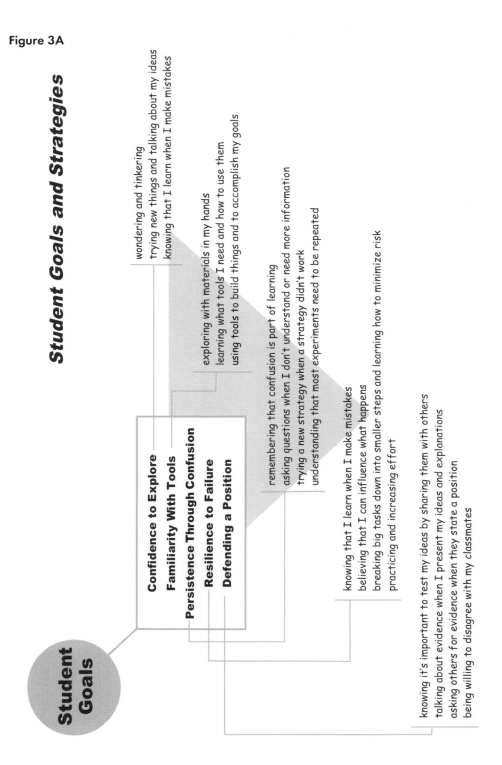

Student Goals and Strategies

Student Goals

- **Confidence to Explore**
- **Familiarity With Tools**
- **Persistence Through Confusion**
- **Resilience to Failure**
- **Defending a Position**

Confidence to Explore
- wondering and tinkering
- trying new things and talking about my ideas
- knowing that I learn when I make mistakes

Familiarity With Tools
- exploring with materials in my hands
- learning what tools I need and how to use them
- using tools to build things and to accomplish my goals

Persistence Through Confusion
- remembering that confusion is part of learning
- asking questions when I don't understand or need more information
- trying a new strategy when a strategy didn't work
- understanding that most experiments need to be repeated

Resilience to Failure
- knowing that I learn when I make mistakes
- believing that I can influence what happens
- breaking big tasks down into smaller steps and learning how to minimize risk
- practicing and increasing effort

Defending a Position
- knowing it's important to test my ideas by sharing them with others
- talking about evidence when I present my ideas and explanations
- asking others for evidence when they state a position
- being willing to disagree with my classmates

Contents

Introduction to the Chapter..**35**

Essay: Confidence to Explore..**37**

    After the Initial Eeewwww 39
    No Longer the Same..41
    Not Having Step-by-Step Instructions................................44
    By the End of the School Year..46
    I Didn't Want to Produce the Same Fears 47

Essay: Familiarity With Tools..**51**

    Don't You Feel Powerful?..54
    The Real Microscope 56
    The Stopwatch as a Tool 57
    Safety Was a Concern 59
    I Don't Even Know How to Use a Saw 61
    The UV Bulb Can Be Changed by the User......................63

Essay: Persistence Through Confusion **67**

    Fun and Frustrating..70
    To Build and Rebuild 72
    Where to Draw the Line 74
    She Wanted to Do It Herself 76

Essay: Resilience to Failure..**79**

    I Shouldn't Have Come 82
    So That All the Bridges Fall 84
    Making Mistakes 86
    Watch Me! 88

Essay: Defending a Position **91**

    I Assumed That Our Girls Would Feel Comfortable..........93
    We Have Reason to Believe 95
    A Little Unnerving..97
    On a More Personal Level 99

Introduction

Triad Student Goals
Confidence to Explore
• wondering and tinkering
• trying new things and talking about my ideas
• knowing that I learn when I make mistakes
Familiarity With Tools
• exploring with materials in my hands
• learning what tools I need and how to use them
• using tools to build things and to accomplish goals
Persistence Through Confusion
• remembering that confusion is part of learning
• asking questions when I don't understand or need more information
• trying a new strategy when a strategy didn't work
• understanding that most experiments need to be repeated
Resilience to Failure
• knowing that I learn when I make mistakes
• believing that I can influence what happens
• breaking big tasks down into smaller steps and learning how to minimize risk
• practicing and increasing effort
Defending a Position
• knowing it's important to test my ideas by sharing them with others
• talking about evidence when I present my ideas and explanations
• asking others for evidence when they state a position
• being willing to disagree with my classmates

The Triad Student Goals—initially born as Girl Goals—were crafted as a tool to aid teachers and scientists in designing science club activities that went beyond fun. In the early years, there was a sense that simply having female scientists, female teachers, and female girls all in the same room doing science together and enjoying it would encourage girls in science. But we came to feel that having fun doing science was not enough. The literature on gender equity in education pointed to evidence that girls are often not given the opportunity to learn skills and habits of mind that are essential to being successful in science. We asked ourselves what traits are characteristic of students who enjoy science and are successful in science. Combining the teaching and scientific expertise from the Triad community with the findings in the literature, the Triad Student Goals were born. What emerged was a picture of a student who is confident in exploring the unknown, who is willing to ask questions, who is not wholly defeated by a wrong turn or a setback. This successful science student tends to persist through problems, confusions, and challenges and takes pride in her own ideas. She also has enough experience and familiarity with tools—both scientific and everyday—to use them confidently. No doubt there are other characteristics that could be listed and added to the Triad Student Goal set; however, these are the starting places that our community used to build strong girls prepared for the challenges and excitement of science. In the following pages, you will find a more detailed consideration of each of the Triad Student Goals—Confidence to Explore, Familiarity with Tools, Persistence Through Confusion, Resilience to Failure, and Defending a Position (See Figure 3A)—highlighting why each of these traits is important for success in science, why girls may have had fewer opportunities to develop these skills, what it might look like to achieve a given goal for a student, and some strategies for how to get there.

SECTION II: Exploring the Triad Framework

Figure 3B

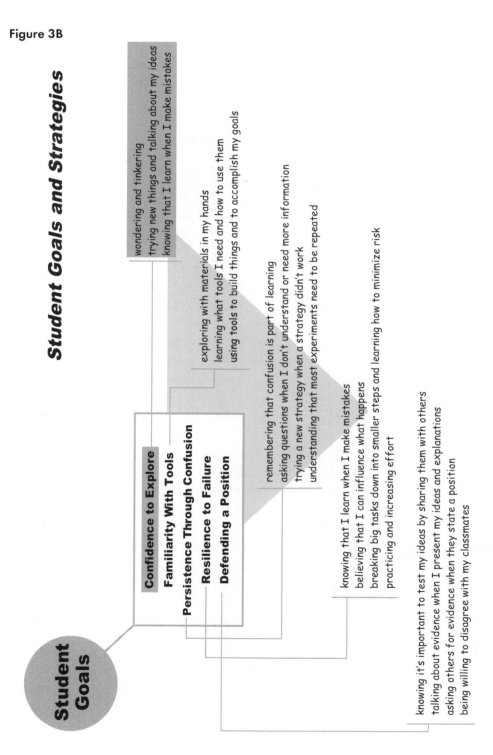

CHAPTER 3: Student Goals

Essay: Confidence to Explore

Fundamentally, science is about exploring and understanding the natural world. The process of being an explorer in science, as in any field, demands confidence: the belief in our ability to succeed through our efforts. Although there are maps to guide us in the form of information that has been drawn up by previous explorers—our fellow scientists—it is confidence that allows us to venture into uncharted waters. We need the self-assurance that we can go forth and return triumphantly. In these uncharted waters we not only discover new things, but we also see old ideas anew through our own eyes. Having the confidence to explore allows us to reinterpret evidence and contribute our own voice and vision to a problem or explanation. Without the confidence needed for exploration, we have little or no chance of making our own discoveries about the natural world around us. Whether we are at the top of our game as a world-renowned scientist or at the beginning as a grade schooler, it is the same act of courage.

In 1989, Marcia Linn and her colleague Janet Hyde performed a meta-analysis of gender differences in adolescence. What they found is that confidence is the principal trait for which gender differences do, in fact, exist (Linn and Hyde 1989). Elizabeth Fenema echoed this finding for girls' confidence in mathematics (AAUW 1998; Fenema 2000). Myra and David Sadker documented that we adults can contribute to girls' lack of confidence by stepping in and doing tasks for them, implicitly reinforcing that we do not believe they can do it on their own (Sadker and Sadker 1994). These rescuing behaviors by adults or more confident peers translate into fewer opportunities for some girls to explore, thereby diminishing their capacity to develop the confidence to be successful in science. Adults' rescuing behaviors and the resulting lowered confidence can lead to a set of behaviors collectively referred to as *learned helplessness* (Peterson et al. 1995). In Judy Gordon's work with Girls, Inc., she explored how a learned-helpless girl will give up at the first sign of difficulty (1995). She will sit back in expectation that the nearby adult or more confident peer will do it for her. She will say in the face of confusion, "I'm not good at that kind of thing," or worse, "I'm stupid." If confidence is key in science, then these self-reinforcing rescuing behaviors by adults in classrooms can result in learned helplessness that have academic consequences that are of particular concern for girls in science.

In science, students cannot gain the confidence to explore without ever having had the opportunity to explore. Learning situations in which students are rescued during an attempt to figure something out, in which there is a single answer expected from

SECTION II: Exploring the Triad Framework

an investigation, in which students most often follow prescribed directions during experiments, or in which unexpected results are ignored rather than celebrated and examined, all reduce opportunities for students to develop the confidence to explore. Learning situations in which students can defer to more confident classmates or get an adult to perform difficult parts of experiments for them do them a disservice.

Our task then as educators is to create learning environments in which students can develop confidence as they engage in science. A few ways to begin to achieve this are to

- provide ample opportunities for girls to investigate freely while pursing science ideas. These experiences become true explorations when they are not externally prescribed by the teacher, a book, or a set of instructions. Much like practicing scientists whose confidence deepens with each discovery made, students' willingness to try out their ideas can grow with each new exploration. This exploration, in turn, can be the basis for both new questions and greater understanding.
- respond to student questions with open-ended questions of your own. "How could we figure that out?" "How could we study it?" "What do you think?" "What do you think will happen if you try that?" This approach can help adults avoid the trap of intellectually rescuing students.
- promote girls' belief that their efforts will result in increased skills and understanding. This is at the heart of confidence. In science, the act of exploration is, in and of itself, a form of success. Giving girls explicit positive feedback on the strategies they try and letting them know what they are doing is what scientists do can help girls develop an attitude of "I don't know what to do next, but I'm confident I can explore more and figure this out." That is the response of a student with the confidence to explore.

CHAPTER 3: Student Goals

After the Initial Eeewwww

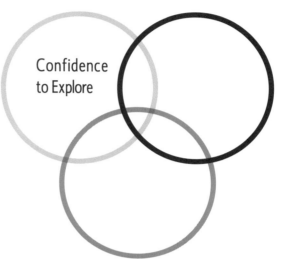

I thought I would discuss the heart and circulatory system on Valentine's Day, which brought on the expected "awws." I was unsure whether I should bring along the real specimens of the heart to display at the end of the club meeting, since I did not know how the kids would react. After getting input from my team members at our planning session, I conducted a straw poll at work among my colleagues (who even gave me input from their kids, similar in age to our club members), all of which persuaded me take along a real heart to the club meeting. The consensus was that after the initial apprehension the kids' natural curiosity would take over, but that we should definitely provide an out for the few who might not to want to touch the specimen. I therefore decided to display the heart and then invite only those who wanted to to touch it at the end of the meeting. I was surprised when after the initial "Eeewwww" everyone apparently wanted to poke their gloved fingers into the specimen, and this taught me yet another lesson.

Reflection Questions

- How did this scientist's expectations of students influence their opportunity to explore during the lesson?
- How do your own expectations as a teacher influence your students' opportunity to explore?
- In what ways did this scientist's lesson actively not encourage students to explore?
- If you were doing this lesson, when would you have brought out the heart? Why?
- What are strategies for structuring lessons that encourage students to explore areas that may at first be intimidating to them? In what other ways might you build students' confidence to explore?

SECTION II: Exploring the Triad Framework

Links

- Student Goals: She Wanted to Do It Herself
- Science Goals: The Real Thing
- Teaching Goals: Keeping Your Hands in Your Pockets
- Teaching Goals: At First I Was Hesitant

CHAPTER 3: Student Goals

No Longer the Same

This vignette is from an observer's notes of the fourth meeting of an all-girls club. There were approximately 17 girls in attendance, more seventh graders than sixth or eighth. They were doing a mouse dissection activity that the girls had requested.

After the students were all seated, the club leaders began to pass out materials. Each student received a dissection tray, rubber gloves, a mask, and a gauze cap. The students seemed excited and chattered nervously. Marlena, a Triad Scientist, went over the utensils in a small kit: scalpel, forceps, scissors, and pins "… to stick the mouse down to the tray." Some girls responded with a characteristic, "E-yuu!" Next Chrystal, the other scientist club sponsor, began an explanation of how to cut. She explained that the diagram was of a rat and that a mouse was even smaller, so they would have to be very careful. She described how to feel for the diaphragm, sternum, and rib cage; how to make the first incision; and how to cut so as to see everything clearly. As Marlena began to place the body of a small white mouse in each girl's tray, Chrystal said, "If you're uncomfortable, spend some time with it first, feel it, move it around. No one has to do it …."

There was a good deal of squeamishness at first, while each girl was waiting to get a mouse placed on her tray. But not long after, when they had their mice, the girls went to work, some asking for assistance pinning them to trays and exclaiming, "Everything's so small!" One group of three girls worked together on a single mouse, but most worked on their own. The students earnestly settled into their task. There was a lot of conversation among themselves and with the scientists. They slowly, cautiously began dissection, breaking the sternum, making the first incision, removing fur. I eavesdropped on random comments:

"Help! Is this okay?"

"We're inside!" "So am I. Is it bleeding?"

Girls in Science

SECTION II: Exploring the Triad Framework

"It's the ribs. You can't cut that."

"Okay, I think that's as deep as I want to go right now...."

"That's the large intestines, see?"

Students continually raised their hands. The noise level reduced as girls work even more intently. Some moments were spent working in silence; others were mixed with frequent outbursts: "Look! I got it!" and "This is cool!" As the work continued and the students acquired more information, their discussions and comments became more inquisitive:

"Is the green thing the intestine?"

"Where's the stomach?" "See. It's right here."

"These are the kidneys." "Do they have two? I can't find the other one."

"It looks like crabmeat."

"Can I see what it ate?"

"I found poop too."

"Oh my, it's [the liver] big!"

"Can we open the tail?"

"What would happen if I cut the heart open?"

"I found a bone behind the heart." Marlena suggested, "Why don't you study the diagram. Does it say what it is? Is it the esophagus? The backbone?"

A group of three girls debated the sex of their mice.

It struck me that by the end of the activity these girls were no longer the same nervous and timid students who began this activity. They had become so comfortable that their energy began to swing to the opposite extreme. Not only were they exploring, some continued cutting, fearlessly separating their mouse in every conceivable fashion—removing eyes and tongue. One girl playfully held up her work calling out, "Look, genuine mouse coat." I didn't think any student was going to walk out of that room quite the same. I knew I wouldn't. That was about as real as discovery can get.

CHAPTER 3: Student Goals

Reflection Questions

- What teaching strategies are the adults in this vignette using to encourage students to explore? What are they doing? What are they not doing?
- In your experience what kinds of science lessons are most successful in building students' confidence to explore?
- Describe a time when you have noticed your own interactions with students encouraging their confidence to explore? limiting their confidence to explore?
- Consider the student statement, "Look, genuine mouse coat." To what extent can a teacher discern and communicate the boundary between exploration in the service of learning and inappropriate use of materials? How might you decide when students' explorations are no longer linked to learning? Is it truly possible for us to know?

Links

- Student Goals: The Real Microscope
- Student Goals: Safety Was a Concern
- Science Goals: Putting Sugar in Water
- Science Goals: The Real Thing
- Science Goals: A Daunting Task
- Teaching Goals: Talking in Questions

SECTION II: Exploring the Triad Framework

Not Having Step-by-Step Instructions

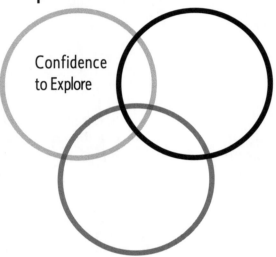

Upon reflection on our club's activities, I was pleased with the progression that the girls made—becoming comfortable with the process of science. We set materials in front of the girls to make lava lamps—an activity designed for them to experiment with liquids of different density. The girls were hesitant to touch the materials at first and were clearly put off by not having step-by-step instructions. Even by the end of that first period, the girls were playing with the materials, asking questions, and performing mini-experiments to see what would happen.

Reflection Questions

- Why do you think these students were hesitant to touch these materials?
- What may have caused them to be put off by the lack of step-by-step instructions?
- When do you think step-by-step instructions limit students' opportunities to explore? When do you think step-by-step instructions are justified?
- What other kinds of supports might be helpful for students in the absence of step-by-step instructions?
- What strategies have you used to transform a traditional lesson with step-by-step instructions into a more open-ended science exploration?
- To what extent are you yourself confident to explore in the absence of step-by-step instructions?

Links

- Student Goals: Fun and Frustrating
- Science Goals: I Learned How a Lava Lamp Works
- Science Goals: Nobody Knows What's Inside

CHAPTER 3: Student Goals

- Science Goals: There Is No Road Map
- Teaching Goals: Keeping Your Hands in Your Pockets

Student Work

3. What was your biggest accomplishment in the Triad Science Club?

I dissected a ~~sheep~~ lamb's heart and I WAS NOT GROSSED OUT! After that I showed my friends the picture from the worksheet and they were all like, "Ew!"

SECTION II: Exploring the Triad Framework

By the End of the School Year

In contrast to the girls' initial hesitation, by the end of the school year they were more than willing to handle and experiment with even the "grossest" of materials. One of our last meetings was dedicated to dissecting fish. My job was to get the girls to touch the fish they were given. I had planned a whole speech about the texture of fish scales to trick the girls into touching the fish. But the speech was totally unnecessary—each and every one of the girls was eager to open the fish to see what was inside.

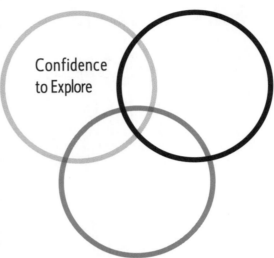

Reflection Questions

- Why do you think this scientist was surprised by her students' response to the fish dissection?
- To what extent do you believe that all students are capable of having the confidence to explore? To what extent do your expectations vary with gender? With other characteristics such as culture, language, and religion?
- What are the critical experiences you think students need to build their confidence to explore?
- Share a story in which you have seen students over time build their confidence to explore.
- This student transformation occurred in a science club. How might you build in opportunities for this kind of change in the science classroom?

Links

- Student Goals: She Wanted to Do It Herself
- Science Goals: Above a Whisper
- Teaching Goals: Talking About Equity

CHAPTER 3: Student Goals

I Didn't Want to Produce the Same Fears

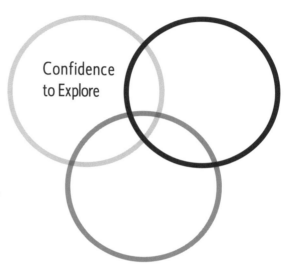

My learning experience with Triad this year has been remarkable. I was engrossed in my personal insecurities. I was a first-year teacher trying to establish my skills. I was enrolled in graduate school, continuing my education in my new field. And I was also participating in a project that involved a subject that I feared the most—science. All of my "isms" arose as I participated with the Triad group. I recalled all of my memories of high school science. I did not understand the subject. I did not fit in with the brain crowd.

As each professional workshop took place, each debrief, each activity, I became less afraid and began to appreciate the fun of learning science. The challenge was to continue to have fun as I worked with our girls in the Girls Science Club. I didn't want to bring my prior isms to our club meetings. I didn't want to produce the same fears that I grew up with in these girls.

Reflection Questions

- How would you describe this teacher's feelings about science? How are they similar to or different from the feelings of other teachers you have known?
- To what extent has your experience of sexism, racism, or other isms been linked to your feelings about an academic subject?
- In what ways might your own feelings about science be perpetuated through your teaching? How specifically might that occur? What does it look like in a classroom?
- If teachers are not confident to explore, what are some strategies they might use to develop their own confidence along with that of their students?

Links

 Student Goals: I Don't Even Know How to Use a Saw

SECTION II: Exploring the Triad Framework

- Student Goals: On a More Personal Level
- Science Goals: Nothing to Do With the Club
- Science Goals: To Trust in Their Own Logic
- Teaching Goals: Anyone But the Boy

Hesitation had disappeared at the end of the year, and every girl wanted to explore the inside of the fish.

SECTION II: Exploring the Triad Framework

Figure 3C

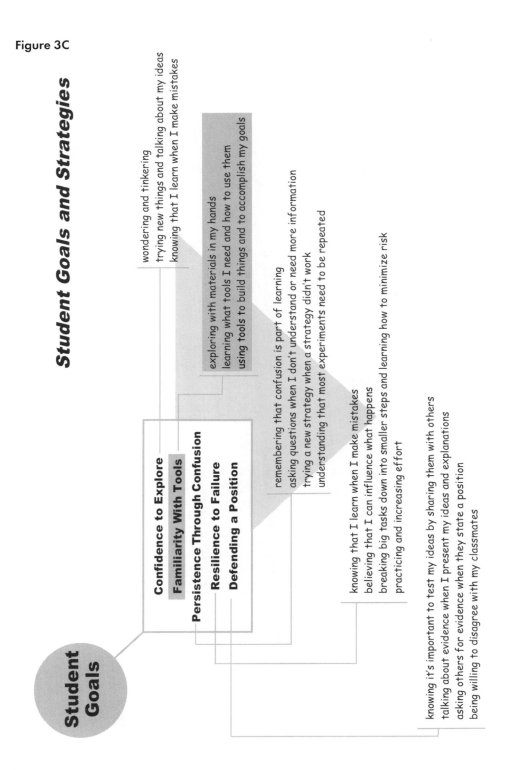

CHAPTER 3: Student Goals

Essay: Familiarity With Tools

From molecules and cells to planets and galaxies, commonly known parts of our natural world could not have been discovered, much less explored, without the development and use of scientific tools. Throughout the scientific disciplines, scientists need tools to extend their senses and to enable them to gather evidence that otherwise would be invisible to them through the naked eye alone or within human time scales. Over the centuries, scientists have developed specialized instruments to aid their quest to understand the natural world. The array of devices that are common to laboratory or fieldwork are endless and range from the simple to the complex: magnifying glasses, telescopes, and microscopes; pickaxes, scalpels, beakers, and test tubes; calculators, computers, and spectrophotometers—and a myriad others so specialized as to be unrecognizable even to other scientists. As they work in field or laboratory settings, scientists must often create custom equipment to aid in the collection of data. These specialized pieces of equipment can run from carrying boxes for insects to supports for telescopes. Most investigative situations require that scientists have familiarity and competence with even the most common, nonspecialized tools such as pliers, wire strippers, knives, saws, hammers, and soldering irons. Sometimes, in resource-limited situations, such as in a remote field investigation or outer space, scientists must improvise, using ordinary materials to perform sophisticated functions.

Familiarity with tools and the ability to use them is essential in science. It's implicit in the rallying call for hands-on, inquiry-based science learning throughout the standards movement of the 1990s. Stanford University researchers Rachel Lotan and Elizabeth Cohen, colleagues during the Triad effort, have found that when students are manipulating materials and talking about the task or the concepts at hand, there is strong correlation with learning (Cohen and Lotan 1997). Tool use is not only important to the authentic practice of science, but also important to learning gains more generally.

Gender socialization is particularly relevant for girls and tool usage. American culture, through avenues such as toys, media, clothing, social interactions, and stereotyping, tends to give messages that aim boys and girls in different directions regarding tool usage. In general, boys are steered away from cooking, sewing, and nursing (all of which also have their own sets of tools and technology), and girls are steered away from construction, automotives, and engineering activities. Girls are often not as encouraged as their male peers in play and activities that involve simple hand tools

SECTION II: Exploring the Triad Framework

such as hammers and nails, saws and vises, wrenches and pliers, nuts and bolts, and wires and batteries. Girls are more often discouraged from engaging in play that involves getting dirty or greasy—which inevitably happens when using tools and building things. As a result, girls may arrive in science classes with less experience manipulating tools and may be more reticent to use them and less comfortable with them in classroom contexts. Further, if boys are perceived as having greater expertise with tools, and girls are reticent to use them, boys will likely have the tools in their hands during classroom science learning.

In Chapter 1, we described a video taken by a team of Triad teachers and scientists of their own students in mixed-sex student groups engaged in a science activity. In that video, some students sat with pieces of tape on their fingers ready for other students to use to make a sail car. All of the carmakers were boys; all of the tape providers were girls. If we are not attentive to the issues of girls and tool use and leave it to chance, we run the grave risk of repeating the essence of that scene. Addressing such situations is complex, and some of the approaches described throughout this book will be applicable to tool usage. Some basic strategies that we have found helpful in encouraging girls' familiarity with tools include the following:

- Make sure that girls have tools in their hands. Obvious but effective. This means not only having tools but also having enough tools for the size of the group. Plan ahead: Think through where there are opportunities for tool usage, count how many tools you have, and figure out ideal group sizes. Carefully assign students to groups, and anticipate which girls may need more time with tools so they can get comfortable with them before they need to use them to complete a task. Consider whether single-sex or mixed-sex groupings may be more appropriate for promoting tool use by all students during the lesson.
- Talk with students about how tools are used in science. Communicate to students that tool use is a learned skill that takes practice and that nobody knows how to use a tool the first time he or she picks it up. Make explicit that different students come to class with different levels of experience with different tools.
- Give girls experiences in which tools extend their strength. Using pulleys to lift 200 pounds is an empowering experience if you weigh 80 pounds! Wrenches with long handles can make stuck bolts turn easily.
- Get comfortable with tools yourself if you aren't already. You are the mentor. If you aren't familiar with tools and comfortable with using them, it will be hard to promote this with girls. If you are a woman working with a male colleague, don't ask him to set up the DVD player for you. Learn how to do it yourself—

CHAPTER 3: Student Goals

and do it. If you are a man working with a female colleague, let her connect the electrodes. One of our best Triad professional development workshops resulted from putting power tools in women's hands and having them build simple hinged boxes out of wooden boards.
- Don't take tools out of girls' hands. A phrase that we found invaluable throughout our Triad experience: Keep your hands in your pockets. When you're talking with a girl who's manipulating materials, keep your hands in your pockets (or clasped behind your back—wherever you can put them that will prevent them from taking over) so that the tools stay in the girls' hands. If you need to show them how to use a tool, go ahead, but then put it right back in their hands and ask them to try what you just did. Just don't do it for them.

SECTION II: Exploring the Triad Framework

Don't You Feel Powerful?

This vignette is from an observer's notes of an all-girls after-school science club for seventh and eighth graders in which the girls were making flutes.

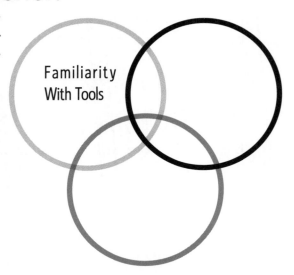

Adreena drilled her holes first, easily. Teresa went next. Teresa was shorter by about eight inches, making it harder for her to get leverage. She struggled for quite a while to make her holes. She couldn't get the drill to bite into the plastic. One scientist suggested that she try a smaller bit to make her initial holes. Teresa switched the bit on her own. The new bit made drilling the holes doable for Teresa. When she was done with the first set of smaller holes she went back over them with the bigger bit. At another table, one scientist helped Sarah put the right bit in the drill. She then gave the drill to Sarah, saying "You're drilling it, not me." Both scientists gave safety advice and also helpful hints on using the drill. As Sarah tried it in the air the first time, her partner Irina said, "Don't you feel powerful?"

Reflection Questions

- How do the adults in this vignette balance letting students struggle and helping them achieve success? What are some additional ways they might do this?
- Describe a situation in which learning to use a tool made you feel powerful. Describe a situation in which not feeling confident with a tool limited your participation in an activity.
- What have been your students' responses to using tools?
- What barriers to student tool use have you experienced? How have or could you overcome them?

Links

 Student Goals: Fun and Frustrating

CHAPTER 3: Student Goals

- Teaching Goals: Can You Help Me?
- Teaching Goals: Keeping Your Hands in Your Pockets

> I was struck by how starved the girls were for using hand and power tools. They were really into it and in some ways even more than with the science equipment. I think of one girl in particular handsawing through a 1 inch by 6 inch board. It was taking her forever, but she refused to use the jigsaw and persevered. It was great!
>
> —*Triad Teacher*

SECTION II: Exploring the Triad Framework

The Real Microscope

I was extremely impressed with the girls' creativity. Without formal instruction, several of the girls figured out how to perform serial dilutions using a pipette. We became aware of how important it was to give them access to the equipment in the lab when one girl remarked, "We feel special when you let us use the real microscope."

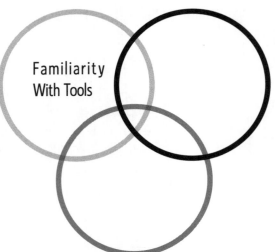

Reflection Questions

- What key insights do you think this scientist had as she observed the girls in her club using scientific tools?
- What about our students' everyday experiences at home and at school might cause them to feel special when they are allowed to use tools like microscopes?
- What kinds of concerns limit your use of tools with your students? How might these concerns be addressed in a way that would give your students increased access to tool use?
- How might you go about helping students to develop skills with tools you are not comfortable using yourself?

Links

- Student Goals: No Longer the Same

- Science Goals: Above a Whisper

- Science Goals: The Real Thing

- Teaching Goals: Can You Help Me?

The Stopwatch as a Tool

During the marble mazes activity, we showed the girls how to use stopwatches. I suspect most of the girls hadn't used stopwatches before, and they loved them. They may have enjoyed the stopwatches as toys, but they didn't really appreciate the use of these instruments for measurement. The girls knew how to start and stop them; but they didn't operate them accurately when timing a marble's travel from release to dropping into the cup. They seemed unfazed by this disconnect between the real travel time and

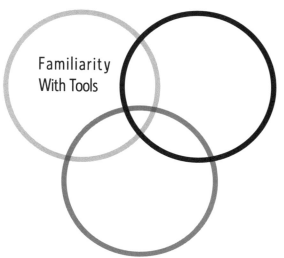

their measurement of it and were pleased to report that their times were getting better when all that was really improving was their ability to run the stopwatch for a desired number of seconds. I think this was important—they were learning how to control the stopwatch. But it didn't have much to do with the marble maze activity. We had intended to direct this activity toward the ideas of precision and accuracy—we asked them to collect three timepoints for the initial and final runs of the marble as a launching point for discussion of these ideas. I think this message ended up being lost on them, because we hadn't taken into account the amount of experience it would take for them to first learn to use the stopwatch as a tool.

Reflection Questions

- What are the key issues this scientist struggled with in her work with students and stopwatches?
- In what ways might limited experiences with tools be a barrier to student learning?
- Describe a time when you have struggled to get your students to focus on precise measurement. How did you address these struggles?
- What kinds of tasks or challenges encourage students to be precise in their tool use?

Links

 Student Goals: We Have Reason to Believe

SECTION II: Exploring the Triad Framework

- Science Goals: Not What We Had Planned
- Teaching Goals: Talking in Questions
- Teaching Goals: Back in the Classroom

In a marble mazes activity, learning to use a stopwatch preempted the planned activity—working on ideas of precision and accuracy.

CHAPTER 3: Student Goals

Safety Was a Concern

I was surprised to find that one of my teammates was extremely uncomfortable with the activity that I felt most strongly about doing: dissecting a cow's eye. It was a major event when we did it in our club last year, and I had said from the start that I was very excited about doing it again. I thought it was absolutely essential to do it, and I knew the kids would find it incredibly fascinating and satisfying. I knew safety was a concern, given the use of scalpels in the activity, but I had full confidence that the children could handle the responsibility if we laid out the rules and provided enough supervision. My teammate, however, felt very nervous about the students using these sharp tools. I thought her concerns were not unfounded but were based on a fearfulness about the students' behavior that I felt didn't really match the behavior of the students in our club. I had the most experience of all my teammates with boys and girls of that age and was comfortable with the kids using scalpels. I didn't want my colleague's fear to prevent the kids from being able to do an exciting activity. I felt we could easily prepare them to handle it well.

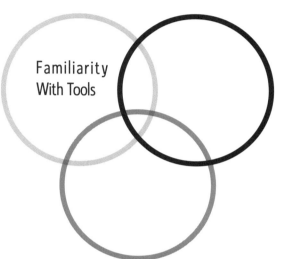

Reflection Questions

- Which of the individuals in the vignette do you identify more with: the author or her colleague? Why?
- Are there tools you would hesitate to use with your students because of safety reasons? What are they, and what are your specific concerns?
- How might you negotiate a middle ground between the two perspectives revealed? What classroom management strategies need to be in place for students to use tools that are potentially dangerous? What might you say to students to instill in them respect and appreciation for tools and their use?
- What attitudes might students develop toward tool use if they consistently receive a message that tools are only for adults?
- Research has shown that hands-on materials in a science classroom most of-

SECTION II: Exploring the Triad Framework

ten end up in the hands of dominant, high-status students. What are teaching strategies for ensuring that all students have the opportunity to use tools, even if they are associated with safety concerns?

Links

- Student Goals: No Longer the Same
- Science Goals: The Real Thing
- Teaching Goals: Keeping Your Hands in Your Pockets
- Teaching Goals: Way Beyond Our Expectations

CHAPTER 3: Student Goals

I Don't Even Know How to Use a Saw

I remember attending the Triad professional development workshop on building wooden boxes. I walked into the lab wearing my teacher attire—long skirt, low-cut boots, and a sensible blouse. As I looked around all the scientists and teachers were sawing, hammering and drilling. I thought, "Oh, no! Wood shop! I can't build! I don't even know how to use a saw!" As always, the very supportive leaders guided me through my first experience of building and at the end of the evening, I created a wooden box. I was so proud.

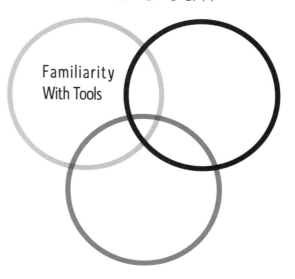

This next vignette is taken from field notes of a science club meeting. A Triad team debriefs as they clean up after a greenhouse building activity:

One of the scientists spoke about an earlier experience that had amused her. She had been sawing extra boards at her lab bench at work in preparation for the club meeting and was surprised how the only two men in the lab immediately jumped in to advise her on how to saw the 'right' way. She later made sure that all the women in the lab got an opportunity to help saw. It turned out that the best sawing work was done by a woman who had never used a saw before.

The following vignette is from a reflection written by a scientist in response to the prompt, "What did you learn in Triad this year and how did you learn these things?

One of the highlights of the club for me, as well as for a lot of the girls, was making boxes. In particular, the first week of box making, which included measuring and sawing, was absolutely amazing. Only a few girls had sawed before, but every one of them mastered it with complete confidence. A few of the girls finished their boxes much faster than everyone else. Once they were done with their own boxes, they helped the other students. They clearly derived a lot of satisfaction from being expert, giving pointers on how to use a screwdriver or hammer. The amount of energy and enthusiasm and pride in that room was fabulous. Although the girls may

Girls in Science

SECTION II: Exploring the Triad Framework

have learned more from other activities in which they persisted through failure and confusion, they gained a lot of confidence in mastering something they didn't know they would be able to do.

Reflection Questions

- What evidence do you see in these vignettes of a connection between general confidence to explore and familiarity with tools? Where have you observed this in your own work or personal life?
- Describe a time when you made something or accomplished a task using a tool you had never used before. How did you feel throughout the process? What helped or would have helped you as you began working with this tool?
- Students come to our classrooms with a variety of experience and expertise. What are some strategies you might use to acknowledge and utilize student expertise while at the same time allowing those students with no experience to develop their own skills?

Links

- Student Goals: I Didn't Want to Produce the Same Fears
- Student Goals: On a More Personal Level
- Science Goals: The Strength of the Group
- Teaching Goals: Can You Help Me?

CHAPTER 3: Student Goals

The UV Bulb Can Be Changed by the User

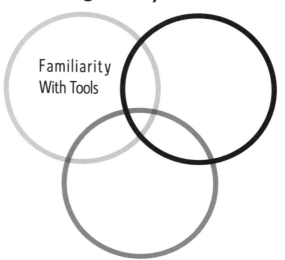

A scientist wrote about an experience in graduate school: While I was a graduate student, the UV bulb on the spectrophotometer burnt out. Since I wanted to use this piece of equipment, I quickly scanned the handbook and found the section on how to change the bulb. The paragraph began with, "The UV bulb can be changed by user." As I was a user of the spec I proceeded to look for the spare UV bulbs. I approached a male colleague and asked him where I could find the spare bulbs. He looked at me and said that I shouldn't touch the spec because I didn't have any experience changing the bulb and that I should just let him do it. Since he was busy at that moment I pointed out to him that the manual said the UV bulb could be changed by the user. He continued to argue with me until I bluntly asked him if he thought I was incompetent! He said that, while there were certainly cases in which I was competent, I was incompetent in this situation. I was angry. I felt he was basing this opinion only on gender stereotypes. Women are not competent with electrical equipment, perhaps. After all, he had no prior knowledge that I personally was incompetent in following instructions. And a part of me was determined to change the bulb in the UV spec regardless of what he said. He didn't own this piece of equipment, and he wasn't in charge. But then I stopped. What if I made a mistake? What if I broke something? I would prove him right! So I left. Defeated. Had I not talked to him at all and become aware of his opinions about me, I would have changed the bulb without fear.

Reflection Questions

- How are the feelings expressed by this author familiar to you? In what ways are they foreign?
- The situation described above about gender interactions is not novel. Teachers and scientists in the Triad community shared countless tales from kindergarten through college and on into their professional lives. Have you observed similar interactions between boys and girls in your classroom? How, if at all, have you responded?

SECTION II: Exploring the Triad Framework

- The young woman in this example fights for herself and then succumbs to her own fear of failure. How might you help students negotiate the conflicting emotions of determination and fear of failure when attempting something new?

Links

- Student Goals: I Don't Even Know How to Use a Saw
- Student Goals: Watch Me!
- Science Goals: Turning to Nadya
- Teaching Goals: Anyone But the Boy
- Teaching Goals: My Own Tendency

Student Work

1. Describe your favorite girls' science activity. Why was it your favorite activity? My favorite activity in girls science this year was making the wooden boxes. This activity was my favorite because I learned to saw, nail, drill and measure wood. I also liked to paint and varnish the wood.

2. What was your biggest accomplishment in the science club? Making my box because I was sooo proud of it and I got to show it of to all my friends and I really liked the way it looked and better yet I MADE IT, all by myself!!(with a little help of my classmates & teachers, but who's to say?)

SECTION II: Exploring the Triad Framework

Figure 3D

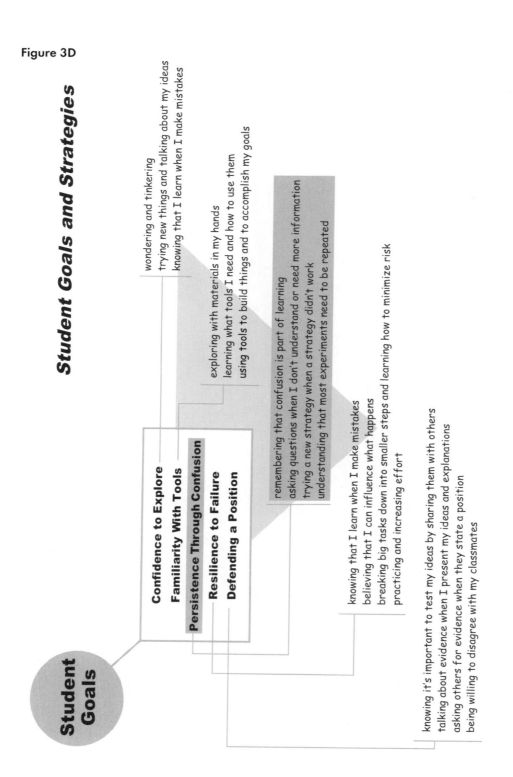

CHAPTER 3: Student Goals

Essay: Persistence Through Confusion

To understand the wonders, the intricacies, and the inner workings of the natural world in all its complexity is confusing. It is confusing to the kindergartner, and it is confusing to the Nobel laureate. It's just a difference of scope and detail. In science, confusion—or incongruence between previous understandings and new findings—is often a signpost on the way to greater understanding. Confusion helps us see the boundaries of our knowledge and raises new questions. Unexpected results in the laboratory are almost always at first confusing to scientists. Persistence through this confusion, however, can be revelatory and lead to new hypotheses, new questions, new experiments, and sometimes paradigm shifts in the way an entire scientific community thinks about the world. In the science classroom, confusion and its resolution are necessary steps toward conceptual change and developing new ideas. Putting students in situations in which they must explicitly confront their prior knowledge and new knowledge has been identified as the critical element in teaching toward understanding (Posner et al. 1982). Persistence through confusion, a willingness to repeatedly attempt to reconcile what appears to be disparate or contradictory information, is a key habit of mind in science and an essential skill in lifelong learning. Persisting through confusion, rather than giving up with a problem unsolved, throwing one's hands in the air, and walking away, is evidence of a strong, confident learner.

However, science and math learning for many students involves confusion along with severe anxiety, fear, and frustration (AAAS 1989). Classroom teachers know all too well the confused, anxious students. They may be highly distracted, acting like the class clown, or they may have a vacant look—the wrinkled brow, dropped jaw, slumped posture, or idle pencil. Then there are the panicked whispers to fellow students, "I don't understand; do you get it?" Teacher attempts to clarify seem only to increase these students' anxieties. Perhaps the following scene resonates with your experience: a bright, lively, girl so overcome with anxiety and confusion over a concept that she sits frozen at her desk, eyes full of tears, unable to make any effort at the task at hand. Any attempts to help her only paralyze her further. Those of us with similar haunting memories might still agonize, asking ourselves, "What could I have done differently? How could I have helped this student to persist through confusion and come out on the other end with her confidence intact?" At its worst, confusion is to find a series of ideas so utterly tangled that one feels overthrown, and this level of confusion, when it happens, can be more devastating to adolescent girls than to boys (AAUW 1998).

SECTION II: Exploring the Triad Framework

In addition, our American society's attitude toward confusion can compound a student's fear and anxiety during learning. So often, society at large views confusion negatively and as a sign of a lack of ability. This attitude can be reinforced in students when they experience science classes in which they are trained to look for clean, simple, singular explanations and are not given the opportunity to wrestle with confusion. In these environments, students may also perceive that asking questions in response to uncertainty or confusion is sign that the person asking the question is dumb.

Students who view confusion as a sign of failure or lack of intelligence are less likely to take on learning challenges. They often avoid the very tasks that will help them understand concepts or develop skills they will need in future learning situations (Dweck 2000). Thus, fear of confusion and of showing our confusion can be a significant impediment to science learning.

If science learning and its inherent confusion can induce such anxiety, then science teaching needs to be structured to allay this anxiety. Herein lies the rub: The source of the anxiety is confusion, but confusion is part of learning. There are many strategies for grappling with this dilemma. Some that the Triad community found particularly helpful are the following:

- Disentangle anxiety and confusion. If girls are to be successful in science learning, we need to support them as they develop their ability and willingness to persist through confusion. Part of this support is to disentangle anxiety from confusion by creating an environment in which the norm is the belief that confusion is part of learning. When we talk together with our students openly and explicitly about confusion as a normal part of learning and a normal part of science, we help to create that norm. "Confusion doesn't mean I'm stupid; it means the material requires effort. If I work hard, I'm going to learn something."
- Identify confusing concepts during planning. Identifying in advance the concepts that will likely be confused by students can be particularly helpful in lesson planning. The *Project 2061 Atlas for Science Literacy Volumes I & II* (AAAS 2001, 2007) is an excellent, comprehensive source for research on student misconceptions relative to concepts, standards, and benchmarks.
- Put students in situations in which latent confusion surfaces. By identifying the concepts that are likely to be confusing, repeated and varied learning situations that evoke latent confusion relative to the concepts can be designed. Lessons are thus structured so that students are intellectually active as they develop functional understanding that resolves conceptual difficulties.

- Assess confusion through asking students to make predictions. Evidence that a student has resolved his or her confusion may be assessed by their ability to accurately predict an outcome. Discussion with peers in this process may serve a number of functions in constructing knowledge. It provides opportunity to reorganize thoughts through talk and listening, and can increase openness to alternative conceptions and sharing ideas (Driver et al. 1994; Duckworth 1996).

SECTION II: Exploring the Triad Framework

Fun and Frustrating

This vignette is from a reflection written by a middle school teacher in response to the prompt, "What did you learn in Triad this year and how did you learn these things?" This teacher included responses from students collected as part of a student evaluation of the club. The evaluation asked the student to draw a picture of their favorite Triad activity and then asked, "Why did you choose to draw that activity?"

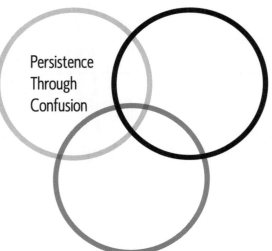

Persistence Through Confusion

One of the students' favorite activities this year was the BB roller coaster. Here are some of the things students wrote about why it was their favorite activity:

The building of the roller coaster took some thought and persistence to decide how the curves and design should go in order for the pellet to move the longest time in the complete coaster.

We were learning about gravity and momentum.

It was fun to make it take as long as possible for the marble to reach the end because we had to experiment with incline and decline of slope.

I know that for it to not go down fast, I had to make it kinda level but steep enough for gravity to still pull it.

I thought this was really fun and frustrating at the same time. It took a lot of improvising and teamwork to make the roller coaster work with all its ups and downs and curves. It was such a relief when it finally worked, and it was the activity that I think I am proudest of.

Reflection Questions

- What do these students' recollections about their favorite Triad club activity reveal about what they value about their science club experiences?

CHAPTER 3: Student Goals

- What is the relationship between fun and frustrating for these students?
- What roles have frustration, confusion, and persistence played in your own experiences as a learner?
- How do you structure learning opportunities in your own classroom so that students experience fun and frustration at the same time?

Links

- Student Goals: Don't You Feel Powerful?

- Student Goals: I Shouldn't Have Come

- Science Goals: Nobody Knows What's Inside

- Science Goals: The Strength of the Group

- Teaching Goals: Does This Bridge Look Better Than It Did the Last Time?

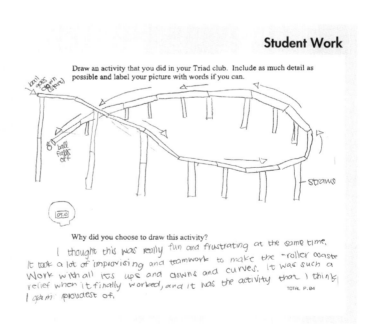

SECTION II: Exploring the Triad Framework

To Build and Rebuild

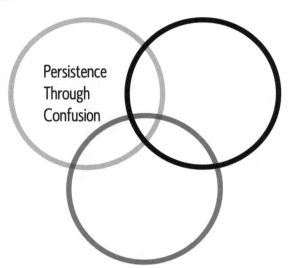

We made moving vehicles that were propelled by mousetraps, or mousetrap cars. Although this activity proved to be materials intensive and was very challenging, it was fun to see the girls try and be persistent through confusion and figure out for themselves how to put the car together so that it would work properly. Their confidence in building based on the diagram seemed stronger than when they built ecosystems in a previous club meeting, but many of the students had to build and rebuild their cars to get them to function properly. As teachers, we encouraged them to help each other more rather than rely on us to help them and encouraged them to figure things out on their own. Many students made modifications to their models to get them to run better, but there were also many who didn't get their cars to run at all.

Reflection Questions

- What does it look like in a classroom when students are "figuring it out on their own"? How is this similar to or different from the science lessons you experienced as a student?
- How does building and rebuilding a mousetrap car build in students the ability to persist through confusion?
- Building and rebuilding takes time. Although it is not always possible to give students multiple chances to try an experiment or investigation, what types of lessons in your own science curriculum might best lend themselves to allowing students multiple tries and building their skills in persisting through confusion?
- Is it okay with you that some students didn't get their cars to run at all? Why or why not?
- How does one ensure that all students are persisting, as opposed to one student doing the activity for their group? What classroom management or other teaching strategies could one use to support all students in persisting through confusion?

CHAPTER 3: Student Goals

Links

- Student Goals: Not Having Step-by-Step Instructions
- Student Goals: So That All the Bridges Fall
- Science Goals: Nobody Knows What's Inside
- Science Goals: There Is No Road Map
- Teaching Goals: Keeping Your Hands in Your Pockets

Where to Draw the Line

I often found myself questioning where to draw the line between making the students struggle through difficult aspects of activities on their own and lending my hands to help them move on to the next step. In almost all cases, however, acting as a sounding board for their thought processes, without contributing much direction at all, was by far the more rewarding path. But, in one activity, the students had to overcome an engineering challenge, part of which involved screwing small screws into blocks of wood. Many of the students could not, by themselves, apply enough pressure at the right angle to push the screw into the wood. I helped several students accomplish this task, either by holding the wood steady, having them hold it while I pushed in the screw, or even working it in by myself. Although the students might have benefited from having a more effective method demonstrated to them, most of them moved on to the next step and had no particular sense of accomplishment in that aspect of the activity.

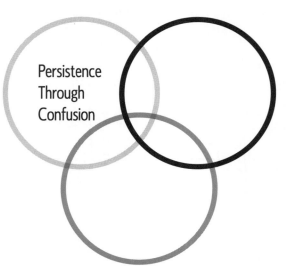

Reflection Questions

- Where did this scientist decide to the "draw the line" with respect to intervening when students encountered confusion or difficulty in their learning?
- How is a student's sense of accomplishment affected by whether or not he or she has to struggle and persist through difficulty?
- In your own experiences as a science learner over the years, where have your science teachers drawn the line in deciding how much help to offer you in your science learning? Share a specific example that you remember.
- Where do you find yourself drawing the line with your own students? To what extent does your line vary with different students and why?
- This vignette describes a science activity in which student persistence is linked to using materials. How might teaching strategies to promote student persistence look similar or different when students are struggling through conceptual confusion?

CHAPTER 3: Student Goals

Links

- Student Goals: Making Mistakes
- Science Goals: The Strength of the Group
- Teaching Goals: Keeping Your Hands in Your Pockets
- Teaching Goals: Can You Help Me?

Physically helping students drive a small screw into a piece of wood may have deprived them of a sense of accomplishment.

SECTION II: Exploring the Triad Framework

She Wanted to Do It Herself

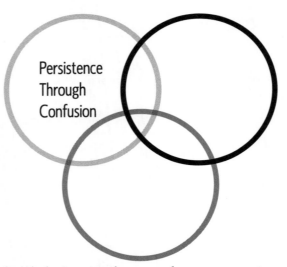

During the first meeting, Lauren was the first girl to ask for help. Something wasn't working, and she needed someone to take over. Her frustration was clear. By the time we did the cow-eye experiment later in the year, Lauren had undergone some fundamental changes. She no longer called for help the minute something didn't work, and her level of frustration was lessened and delayed. I remember watching as she attempted to pull the retina from the back of the eyeball. She was having significant trouble. Noticing her difficulty, one of her friends offered to help Lauren, but, much to my surprise, Lauren said she wanted to do it herself.

Reflection Questions

- What school science learning experiences might have contributed to Lauren's (initially) being the first girl to ask for help?
- Research literature on gender equity has described *rescuing*—the tendency for adults to step in and do difficult tasks for girls—as contributing to their unwillingness to persist through difficulty and confusion in science. To what extent have you observed this in adult-student interactions?
- What are concrete teaching strategies for allowing students like Lauren to persist through difficulty and confusion? At what point would you step in and "rescue" a student like Lauren during her frustration?
- What might have been critical learning experiences for Lauren (or other students you know like her) to get her to the place where "she wanted to do it herself"?
- In what ways can we structure group work in classrooms so that students like Lauren have the chance to do it themselves without other group members taking over?

CHAPTER 3: Student Goals

Links

- Student Goals: By the End of the School Year
- Science Goals: The Strength of the Group
- Teaching Goals: I Watched in Awe
- Teaching Goals: Many People Got a Chance

> I was visiting a Triad girls science club one afternoon when I saw an interaction unfold that beautifully illustrated the real need for this type of environment for girls. The club was making kites that day, and the girl I was watching was using a utility knife to cut a piece of dowel. She was struggling a bit but was making headway. Her brother was sitting at the table with her, waiting to walk her home. He was watching. Then he was twitching. You can probably predict what happened next. First he started to make fun of her, joking that she was about to lose control of the knife, which would go winging across the room and land on the teacher. They both laughed at this. Then he said, "Give me that," and she did. He started to work on cutting the piece of dowel, struggling a bit, making some headway. At this point I was uncertain how to handle the situation so I called the teacher over. She took one look at this scenario and demanded, "What's going on here?" The brother said, "She can't do it." By this time there were four adults gathered around the table and there was a quadraphonic chorus of "Oh, yes she can!!!" The brother waited out in the hall, the sister finished her kite, and I came away knowing that the Triad Project was giving these girls the opportunity to struggle and to succeed or fail on their own terms, where no one could say "she can't do it" and get away with it!

SECTION II: Exploring the Triad Framework

Figure 3E

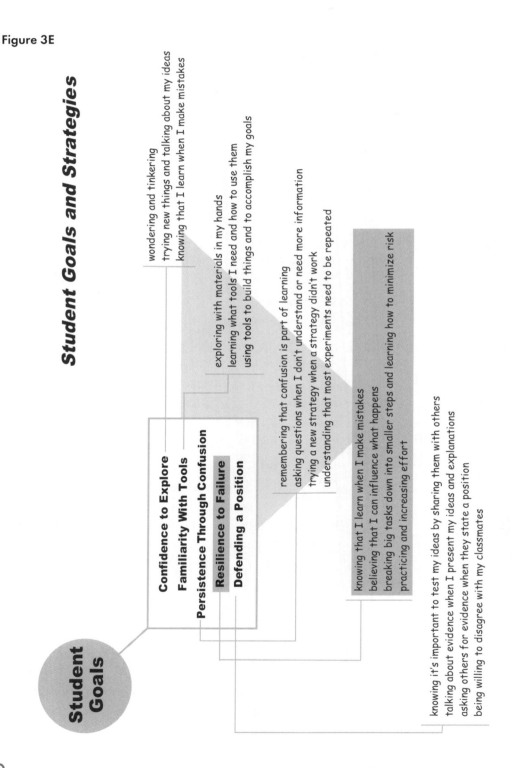

CHAPTER 3: Student Goals

Essay: Resilience to Failure

Over and over in our work with and as scientists, we've heard the phrase, "99% of experiments fail." More often than not, in the scientific laboratory an experiment doesn't work—it "fails." An experiment can "fail" when technical aspects don't work, or it can "fail" when the experiment yields an outcome that wasn't predicted and thus doesn't yield evidence for a particular hypothesis. The first kind of failure, technical failure, happens because investigations and experiments are by their very nature messy. It is no simple matter to frame a testable question and design and carry out a series of experiments that disentangle all of the relevant variables, allowing one to capture meaningful results. In fact, it is the rare experiment that works cleanly the first time. A scientist must be willing to analyze, troubleshoot, and try again. The second kind of failure, failure due to an unexpected outcome, is perhaps an inappropriate use of the term *fail*. Unless technically flawed, most experiments still provide useful information and insights to scientists even if the results are not what were originally predicted. Unexpected results are an essential aspect of investigation, and these "failures" often provide key insights. They spur new questions and drive entirely novel lines of scientific inquiry. Some scientists would argue that the unexpected outcome is of more value than the expected one. Often, it is resilience in the face of these unexpected results that pushes a scientist forward in the direction of great discoveries.

This notion of failure is critical in science, and it is critical for girls because girls, more than boys, tend to believe intelligence is innate and fixed over time, tend to choose tasks that present little challenge, and tend to internalize failure (Dweck and Leggett 1988; Licht and Dweck 1984). Internalization of failure can affect a girl's willingness to engage in scientific investigations where experiments need to be repeated, where they involve uncertainty, and where she could be perceived as "failing." Educational researcher Mary Budd Rowe found differences between individuals as to how they internalize success or failure and found strong relationships between this internalization and an individual's persistence and confidence (Rowe 1974). We can believe that we have personal responsibility for events, or we can believe that they are beyond our control. If you believe that you can influence events—namely, what you do matters—you will tend to have higher motivation. Conversely, if you believe that you can't influence what happens—what you do doesn't really matter—you will tend to have lower motivation. Rowe called this *fate control*. "What you do matters" is high fate control; "what you do doesn't matter" is low fate control. Many studies indicate that males tend to have a higher fate control than females (Nielsen and Long 1981; Chubb 1997).

SECTION II: Exploring the Triad Framework

Confident individuals, those with *high fate control*, not only believe that they influence results but also have the skills to evaluate outcomes. They are more likely to assign responsibility for failure to external forces—such as technical problems and equipment failure—and less likely to feel that the failure was due to their own lack of abilities. These confident individuals are also more likely to assign success to themselves; they succeeded because they were smart, they worked hard, or they did a good job.

Conversely, unsure individuals, those with *low fate control*, not only believe that they don't influence results but are also more likely to assign responsibility for failure to themselves because they believe they are stupid, unprepared, or not good at whatever they happen to be attempting. These nonconfident individuals are likely to assign success to an external force: They had good luck, or the task must have been easy. It's a double whammy: "If I do well, it was luck. If I don't do well, it's my fault." These individuals have low task persistence and often give up at the first sign of difficulty. With poor skills in evaluating consequences, they also have trouble with risk analysis. These individuals may have difficulty, for example, breaking a complex task into smaller pieces to reduce the risk at each step. They also are more likely to choose tasks that have a high likelihood of success and avoid more open-ended tasks for which the outcome is unknown. In keeping with this pattern, they strongly favor immediate reward rather than a larger but delayed reward. They are not resilient.

It is essential for students of science to embrace unexpected results and technical challenges as part of the enterprise of science and not a reflection of personal attributes. It is critical that they see that failures are a part of learning and that multiple attempts in experimentation are the norm for even the most successful of scientists. Some strategies that the Triad community has found useful in developing resiliency to failure in students include the following:

- Talking with students about how it's normal in science for experiments to "fail." Consider reframing the term *failure* to mean *unanticipated results*. Work with students to understand that, in science, one learns from investigations regardless of the outcome. One can learn in many aspects: how to rephrase a question; how to improve a technical setup; how the world works or doesn't work.
- Having girls spend time analyzing results to see that they can influence what happens. Provide ample opportunity for analysis of why unexpected results occur sometimes in science and how these results can inform future experiments. If time isn't spent in this phase, how can one see that she has control over the experiment and that it's not just fate? Take the time to show her that what she does does indeed matter.

- Teaching appropriate risk-taking strategies. Girls need guidance on how to reduce risk and how to learn from mistakes. Strategies on how to avoid technical failures can include thought experiments, making predications about possible outcomes, setting up simple versions of more complex experiments, and interpreting what those results could mean prior to physically setting up the complex version. Students and teachers can discuss with one another explicitly that making multiple attempts at designing an experiment, building a model, or coming up with a question to investigate is okay and something that scientists do in their laboratories every day.

SECTION II: Exploring the Triad Framework

I Shouldn't Have Come

This vignette is from field notes of a science club meeting. Students built roller coasters for metal pellets, or BBs, out of drinking straws. They were challenged to build a coaster that made the BB take the longest possible time to pass through, using a limited number of straws.

Kelland says—almost stomping her foot—"This is so hard I shouldn't have come." A few minutes later, Kelland is talking about this activity being dumb. The next time I look over, Kelland is hitting her tube where the BB got stuck in it. Kelland and Christina have a steep drop going to the ground, but are having trouble making it stick to the table.

Reflection Questions

- Have you encountered a student like Kelland? Describe a student who comes to mind.
- What would you say to Kelland to keep her from giving up on the activity?
- Research literature has described that girls are less likely to take risks that may result in failure. To what extent have you observed this in your own teaching?
- What are concrete teaching strategies for encouraging students like Kelland to take risks that sometimes will result in failure?
- How might Kelland's resistance to risk taking ultimately affect her opportunities to learn?

Links

Student Goals: Fun and Frustrating

Science Goals: Turning to Nadya

Teaching Goals: I Could See How Much I Learned

CHAPTER 3: Student Goals

> Frustration and confusion are signs of learning—necessary parts of learning. When your hypothesis is wrong and you have to do a million more experiments than you had hoped, you learn more than had it been right (as I was often told and finally experienced a couple of years ago). When your hypothesis is right, all you do is confirm.
>
> —*Triad Teacher*

SECTION II: Exploring the Triad Framework

So That All the Bridges Fall

This vignette is from a reflection by a Triad scientist about her team's planning discussion on how to structure a bridge building activity to address the Triad student goal of Resilience to Failure.

A key thing I learned during my Triad experiences was that if I wanted students to become resilient to failure, then not only did I have to give them opportunities to fail, but I also actually had to make them fail. This seemed counterintuitive at first. If I want my students to be more resilient, how is experiencing failure going to get them there? But I had this insight during a wonderful conversation with my teammates while planning for a club meeting. We had done dissections and purifying DNA and all sorts of biology lessons, and now we wanted to give our students a chance to become engineers. We had settled on bridge building with straws and pins as a cheap, easy engineering activity, which was relevant to our students because we live in an area with lots of bridges. Our discussion was all about how to do this bridge building activity. If we charged the students with building a bridge that would hold the most weight, then we'd set up competition and there would end up being one winning team. If there were no weight challenge, then it would just be about building pretty bridges with no engineering and little science. We were discussing these two approaches when my scientist partner asked the question, "Well, what student goal are we really trying to address?" Then it all came together. We agreed that this lesson was about building students' resilience to failure, and thus all of the students needed to fail in their first attempts to make the bridge. None of the first bridges we adults had built during our planning meeting could hold four toy cars, but a couple of our revised bridges could. We decided to challenge all the teams to build a bridge that would hold at least four toy cars so that all the bridges would fall on their first test. During the club meeting, none of our students built a bridge that could hold four toy cars on the first try, and all but one team eventually had built a bridge that could.

CHAPTER 3: Student Goals

Reflection Questions

- How did the student goal of Resilience to Failure guide this team's decisions about how to structure their science club lesson?
- What insight did this scientist have into the role of failure in learning?
- In your own teaching, when do students have opportunities to fail? When, if ever, do you intentionally engage students in an activity in which they will have to experience failure?
- How would you have talked with the one student team whose bridge never held four toy cars, the team that "failed"?

Links

- Student Goals: To Build and Rebuild

- Science Goals: Nobody Knows What's Inside

- Teaching Goals: Does This Bridge Look Better Than It Did Last Time?

- Teaching Goals: Personal Development

SECTION II: Exploring the Triad Framework

Making Mistakes

Working through mistakes, when we as teachers encountered unexpected difficulties, was another powerful learning experience. In preparing for one activity about making soap, we mixed the lye in bowls that turned out to be aluminum alloy. What resulted was a corrosive disaster that turned into a great visual aid and lesson for the whole group. We extended the activity into the next week and took some time to discuss safety issues and precautions, our own accident a prime example of things that could go wrong. I imagine it was useful, in the long run, for students to see teachers and scientists making mistakes and working through them.

Reflection Questions

- Why did these teachers and scientists decide to share the mistakes they made with their students?
- Describe a science teaching experience in which things went awry. How did you handle it? What did you tell your students? What did you not tell them and why?
- Does sharing our mistakes with students compromise our authority in the classroom? Why or why not?
- Although mistakes are accepted as a normal part of the learning process—especially in the context of scientific inquiry—students are rarely rewarded for making mistakes, acknowledging them, and being resilient and trying again. How might one actually reward students who take risks, make mistakes, and keep trying in the context of a science class?

Links

Student Goals: I Didn't Want to Produce the Same Fears

Science Goals: Scientifically Dissatisfied

CHAPTER 3: Student Goals

- Teaching Goals: Stop in My Tracks
- Teaching Goals: My Own Tendency

SECTION II: Exploring the Triad Framework

Watch Me!

This vignette is from field notes written during an observation of an all-girls Triad science club.

The meeting began with snack and attendance. Sonia, one of the Triad scientists, reminded the girls that last club meeting they had talked as a group about why they were having a girls science club. She said that this time she and Kara would talk a little bit about why they became scientists. She asked the girls for their ideas about why they thought they'd become scientists. Answers included: "to learn more," there were "too many male scientists," they wanted "to do science or liked science," they wanted "to explore the world," they wanted "to know how to do experiments," and they wanted "to help kids."

Kara explained that she became a scientist because she liked science, wanted to learn more about the world, and liked to figure things out. One of the teachers asked her to say something about the science teachers she'd had. Kara talked about her fifth-grade teacher who was really exciting and had won awards. Kara had made cameras and dissected a cow's eye in his class, and it had gotten her excited about science. One of the teachers asked if it was gross to dissect a cow's eye. Kara said it was gross, but fun. The whole room squealed and eeewwwwed.

Sonia then described how she never planned to be a scientist. She wanted to be an artist when she was growing up. She also described a "terrible experience" in eighth-grade science with a teacher who was mean to her and wrote a note that got sent home to her parents saying Sonia could never do science. She said this made her angry, so she wrote him a letter back and said, "Watch me! I'll do well in science." She said that she had good science teachers in high school.

Several girls asked follow-up questions about Sonia's experience with the mean teacher. They were very intrigued by the story, and one girl eventually asked Kara if she had had a mean science teacher like Sonia had. Kara said, yes, she had, in college and that her college teacher had told a room full of 200 students that they were all stupid and no good and shouldn't be in the class.

CHAPTER 3: Student Goals

Reflection Questions

- How are the stories that these two women share about their pathways to science similar? How are their stories different?
- Sonia tells a story of how a teacher predicted that she would fail in science. How is the prediction of failure by a teacher different from actually experiencing failure oneself? For you, which seems worse?
- Have you had someone tell you that you were not good at something or that you could not be successful in a particular academic subject? How did this feel? How did you react in the short term? In the long term?
- "It's not rocket science" and "It's not brain surgery" are common statements. Why do you think there is a general perception that only some people can succeed at science and most will fail?
- What are strategies for building strong students who are both resilient to experiences of failure and defiant in the face of those who would predict their failure in science?

Links

- Student Goals: The UV Bulb Can Be Changed By the User
- Science Goals: Walking Encyclopedia
- Science Goals: When Science First Was Really New
- Teaching Goals: Like Dad
- Teaching Goals: Talking About Equity

Student Work

SECTION II: Exploring the Triad Framework

Figure 3F

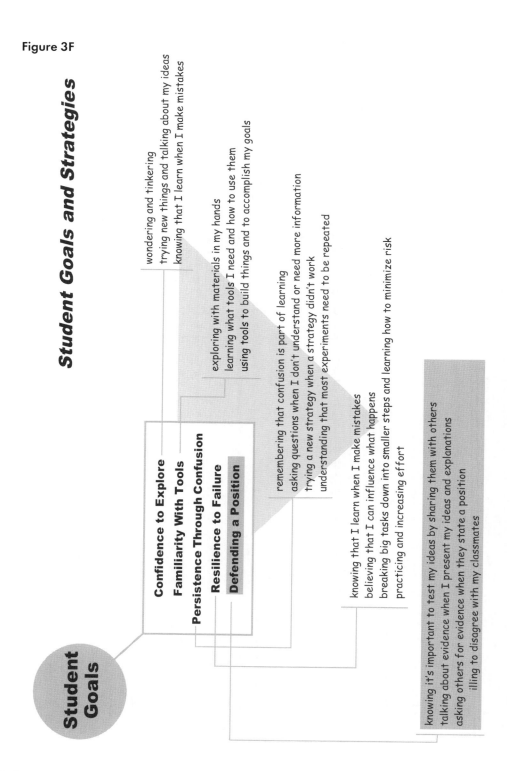

CHAPTER 3: Student Goals

Essay: Defending a Position

Rigorous debate is the norm in the professional culture of science; it fuels forward progress. Everyday, scientists gather all around the world to share evidence with one another, defend their current explanations of natural phenomena, and hash out common understandings. In order to function within this scientific community, scientists must be able to state clearly their ideas and assertions, describe scientific evidence, and articulate the logic and reasoning. Talking about evidence, asking others questions about their lines of logic and the interpretation of evidence, and being scrutinized about your own evidence are all part of scientific discourse. In this environment, scientists must be confident enough to put their ideas out in the world and to endure the scrutiny and questioning of their colleagues. It's also duty of scientists to question their colleagues' work in the same manner. Defending positions in this manner is at the heart of scientific habits of mind. Some of the most important breakthroughs in science have come when a researcher has endured others' intense scrutiny and skepticism by refining and expanding a strong argument through valid reasoning and the gathering of evidence. At other times, scientists must be willing to change their positions regarding long-held explanations upon seeing that they no longer hold up in light of new evidence and analysis.

Although this scrutiny and questioning can appear like arguing in a general sense—disagreeing with someone—it has a more precise meaning in science and is key to the discipline of science. An argument, in the scientific sense, is a coherent set of statements that move from a premise to a conclusion through the use of evidence and reasoning. As such, arguments can be valid or flawed. *The Atlas of Scientific Literacy, Vol. II,* organizes detecting flaws in arguments into seven strands: detecting bias, detecting misuse of numbers, detecting overgeneralization, detecting unfair comparisons, detecting flawed reasoning, detecting alternative explanations, and detecting unsupported claims (AAAS 2007). These argumentation skills work for analyzing and defending one's own or another's argument.

The act of defending a position takes more than confidence: It takes courage. Our interactions with girls often are at odds with their development of the skills needed to take a position, defend it with evidence and logic, and withstand its criticism. These interaction patterns can be related to adult behaviors that foster learned helplessness (see the essay on Confidence to Explore, p. 37) or from our socializing of girls with respect to discourse. For example, although girls are less likely to interrupt someone than boys are, if they do interrupt, we are more likely to reprimand them. We're more

SECTION II: Exploring the Triad Framework

likely to remove a boy from conflict whereas girls are expected to resolve a conflict. We tend to give more substantive and critical feedback to boys than to girls, to whom we tend to give feedback of a nonspecific and nonintellectual nature. Adolescent girls in particular may be subject to images of women presented by the media or society as a whole that suggest passivity is more attractive, appropriate, and acceptable in social interactions. To help students develop their abilities, you could try the following:

- Help students understand that communication in science is about testing ideas, not just sharing them. So often, student presentations or reports are a one-way street and intended for the teacher, not for fellow students. If students, however, are all exploring different lines of inquiry around a central concept, questions, comparisons, and the development and revision of shared ideas can emerge. Testing ideas with fellow students can also help to create a classroom culture in which students can respectfully disagree with or question positions.
- Work with students to connect evidence and reasoning to explanations. The benchmarks in the "Detecting Flaws in Arguments" map in the *Atlas for Science Literacy, Vol. II* (AAAS 2007) is a concise place to find concepts associated with logic and reasoning, which in turn can be explored with students.
- Use many modes to communicate science findings. The four different modes of mathematical, visual, oral, and written communication can help students analyze and understand findings more fully, which, in turn, provides a basis for confidence.
- Link to the Science Goals essays in Chapter 4 on "Use Evidence to Predict, Explain, and Model" and "Think Critically, Logically, and Skeptically".

I Assumed That Our Girls Would Feel Comfortable

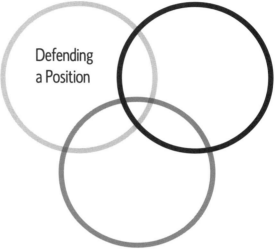

I assumed that our girls would feel comfortable and at ease learning science. I thought that, without the presence of boys to compete against or impress, the girls would naturally blossom and feel comfortable in the club setting. I was mistaken. Many girls did not share as freely as I had hoped for. The equity concept was not limited just to the presence of boys but also to many other issues, such as second-language learners or public-speaking fears or lack of comprehension of our assigned lesson. We, the three scientists and two educators, debriefed on which students were more outspoken, which ones grasped science concepts easily and quickly, which ones were not inhibited in voicing their confusion, and which ones sat quiet in the back.

Reflection Questions

- Why do you think this teacher was surprised that her students were not as comfortable sharing their ideas as she had expected?
- What issues does this teacher identify that seemed to contribute to her students' reluctance to voice their ideas? What additional issues have you observed in your own classroom?
- Do you think it is important that every student learn to share and defend his or her position? Why or why not? What are the potential long-term consequences for a student who does not develop this skill?
- What potential strategies might you use to engage every student in learning to voice and defend a position?

Links

Student Goals: I Shouldn't Have Come

Science Goals: Turning to Nadya

SECTION II: Exploring the Triad Framework

- Teaching Goals: I Have to Introduce Triad
- Teaching Goals: By Scoring When a Girl Participated
- Teaching Goals: Many People Got a Chance

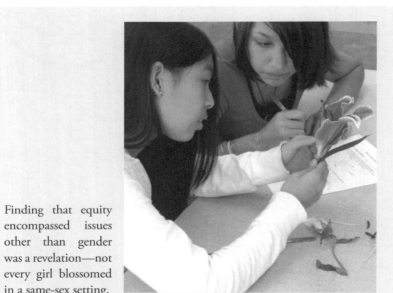

Finding that equity encompassed issues other than gender was a revelation—not every girl blossomed in a same-sex setting.

We Have Reason to Believe

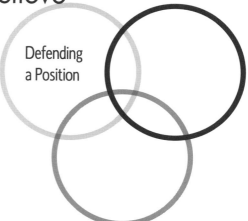

This vignette is from an observer's audiotape of an all-girls club at a middle school, where the girls used different types of forensic analyses of samples found at a crime scene—such as hair, fingerprints, powders, and the ink from a note—to build an argument as to which of four suspects scrawled "Science Club Rules" across the playground wall. The four suspects discussed are referred to as: Suspect A, Suspect B, Suspect C, and Suspect D. The assigned reporters from each of five groups are taking turns reporting on their group's conclusions.

Sun Yung: We found on the chromatography thing that we thought it could be Suspect D, but then it's like darker on the bottom and then it's lighter and then it's blue. You have blue on Suspect A's, and it's darker, and Suspect D's [pen chromatography] is a lot lighter. And, uh, yeah …

Janine: Suspect A's index finger matches this one [that was found at the crime scene], and we also looked at it under a magnifying glass and then, uh …

Ilene: We have reason to believe that Suspect C did it because her hair.… And there's some white hair in there [girls laugh] so we can kind of assume from that and then the fingerprints—

Kelly: The fingerprints—I think that it was Suspect C's fingerprints because they seemed to match the things [found at the crime scene], but it might've been Suspect B's, but it's hard to tell …

Sun Yung: Okay, now the powders. The powders for Suspect C were the exact [same as the crime scene]. They looked like they were the same color, and they were like the same coarseness. And like they both like sank to the bottom when you put water in it, and it didn't rise up like some of them. Then, when you mixed it, the pH were [sic] almost the same. And then we looked at it, and the way they looked they were both kind of clumpy when you poured water on it.

Sarah: Okay, we had it narrowed down to Suspect C and Suspect A. We looked at it closer, and we ruled out Suspect A.

Sun Yung: And because there's a line from the [pen chromatography of the] note that's kinda more reddish in the middle. And Suspect C had more red than Suspect A …

SECTION II: Exploring the Triad Framework

Reflection Questions

- What evidence do you see in this vignette of students defending a position? What additional kinds of talk might you wish to see in this discussion among students?
- Are there examples above of students' making conclusions without using evidence to support their ideas? What kinds of questions or prompts do you think that the teachers and scientists could have used to push students to use evidence in their arguments?
- What do you notice about how the conclusions of the students evolve over the course of their discussion?
- What strategies have you found helpful in keeping students focused on evidence they have collected as opposed to their prior knowledge, opinions, or feelings?

Links

- Student Goals: The Stopwatch as a Tool
- Science Goals: Walking Encyclopedia
- Science Goals: You Can Lead a Horse to Water
- Teaching Goals: I Have to Introduce Triad
- Teaching Goals: Many People Got a Chance

Student Work

> In one Triad club we had to solve a mystery. Our counselor came in and told us a crime had been committed and the suspects were the scientists and the teachers leading the club. We had to look at hair, powders, fingerprints, and ink. It was really fun to try and find out who committed the crime. We learned all about the forensic science police use to solve crimes.

CHAPTER 3: Student Goals

A Little Unnerving

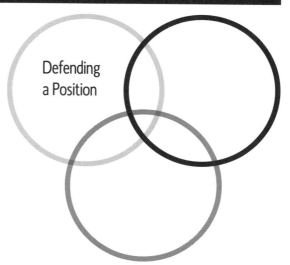

After we (my assigned partner and I) did the pendulum experiment in one workshop, we plotted the results on a graph. Because the graph had a title and the axes labeled, it seemed self-explanatory. However, the teachers in the discussion group suggested we explain what our conclusions were in words; why we had chosen the x-axis to plot the pendulum length and the y-axis to plot the duration of the swing, and why our graph did not, at least at first, appear to represent the point we were apparently trying to make. Though this was a little unnerving (I had a "how dare you question the obvious" response), we did manage to defend our results successfully, and this stood me in very good stead when I had the children plot the result of their taste experiment at a club meeting later in the year.

Reflection Questions

- Why do you think this scientist found the suggestion to explain his graph of his data "unnerving"?
- What is the potential value of asking someone to explain something they may see as self-explanatory?
- As a learner, what is the difference between being pushed to explain by fellow learners as opposed to being questioned by the teacher? Who might benefit and why?
- Do you think questioning the obvious is an important part of defending a position? Why or why not?

Links

- Student Goals: Making Mistakes
- Science Goals: You Can Lead a Horse to Water
- Teaching Goals: Talking in Questions
- Teaching Goals: The Quieter Girls

Girls in Science

SECTION II: Exploring the Triad Framework

> **Pendulums: Modeling a Rope Swing**
>
> Students use pendulums to model the behavior of a rope swing. Students first build their own pendulum, play with it some, then share questions or problems that came up. Through this class discussion, the variables of length, weight, and drop angle are identified. In pairs, students then collect data to determine the effects of these variables on the number of swings in a given unit of time. Students discuss, report, and/or write about the implications of their results for the design of a playground rope swing.

On a More Personal Level

This vignette is from a reflection written by a middle school teacher. At her school site, this teacher worked primarily with special-education students during the regular school day.

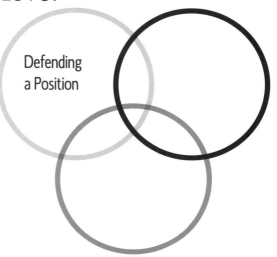

On a more personal level, I was reminded that I am not very good at standing up for myself in stressful situations. My site administrator (thankfully not my boss) tended to view me as overflow staff. She really didn't see the benefit of what I teach the kids. I might be much happier at my campus if I would take the time to convince her that what I teach the kids is important and effective. So, when she told me that she was "really happy that I was willing to help out with escorting children home for the after-school science club," I felt pretty undervalued. And I didn't remind her of what my role in the science club was. I just smiled. I needed to work on that as my own personal Girl Goal—defending a position (specifically mine). I discussed this with another Triad team member. Through my discussion with her and my resolve to do a better job of defending a position with data, I made an appointment with my site administrator to show her beginning/ending test scores of my students. It was a start.

Reflection Questions

- How does this teacher appear to be approaching the problem of her site administrator's perceptions of her value? To what extent do you think awareness and language to describe the problem were helpful to this teacher?
- Describe a situation in which you wished you were better at defending a position.
- What strategies might you use as a teacher to help students build the skills and confidence to defend their positions?
- Why do you think this teacher chooses to add the phrase "with data" to defending a position?
- Do you think the use of data is an important part of defending a position? Why or why not?
- Is the value of data in defending a position limited to science? What strategies might you use to encourage students to use data in defending their positions in subjects other than science?

SECTION II: Exploring the Triad Framework

Links

- Student Goals: I Don't Even Know How to Use a Saw
- Student Goals: The UV Bulb Can Be Changed by the User
- Science Goals: Nothing to Do With the Club
- Teaching Goals: My Own Tendency
- Teaching Goals: Personal Development

Sometimes teachers can benefit from their own "girl goals"—in this case, Defending a Position.

CHAPTER 4:
Science Goals—Envisioning Science in Classrooms

SECTION II: Exploring the Triad Framework

Figure 4A

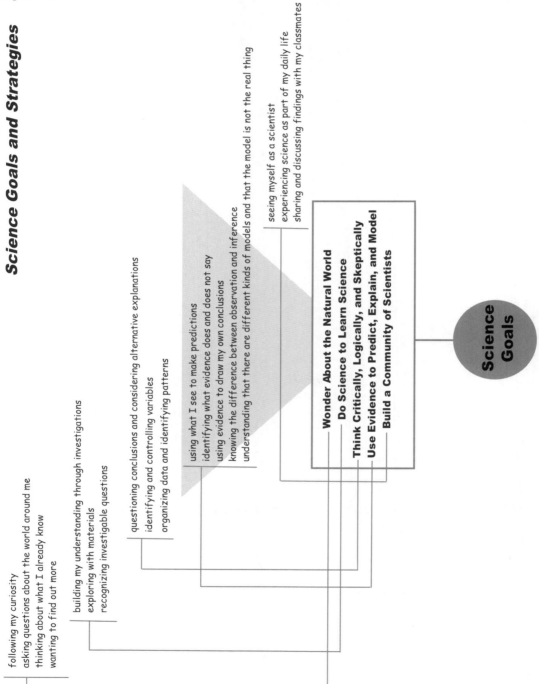

Science Goals Table of Contents

Introduction to the Chapter **105**

Essay: Wonder About the Natural World **107**

    Not What We Had Planned 109
    To Simply Marvel 111
    The Balloon Droops 113
    Above a Whisper 115
    Nothing to Do With the Club 117

Essay: Do Science to Learn Science **121**

    Putting Sugar in Water 123
    I Learned How a Lava Lamp Works 125
    Nobody Knows What's Inside 127
    The Real Thing 129
    A Daunting Task 131

Essay: Think Critically, Logically, and Skeptically **135**

    Answers Are Not the Goals 138
    Walking Encyclopedia 140
    To Trust in Their Own Logic 142
    There Is No Road Map 144

Essay: Use Evidence to Predict, Explain, and Model **147**

    The Most Difficult 150
    Scientifically Dissatisfied 152
    You Can Lead a Horse to Water 154

Essay: Build a Community of Scientists **157**

    The Strength of the Group 159
    When Science First Was Really New 161
    Have to Make It Right 163
    To Help and Teach Each Other 165
    Turning to Nadya 167

Introduction

Triad Science Goals
Wonder About the Natural World
• following my curiosity
• asking questions about the world around me
• thinking about what I already know
• wanting to find out more
Do Science to Learn Science
• building my understanding through investigations
• exploring with materials
• recognizing investigable questions
Think Critically, Logically, and Skeptically
• questioning conclusions and considering alternative explanations
• identifying and controlling variables
• organizing data and identifying patterns
Use Evidence to Predict, Explain, and Model
• using what I see to make predictions
• identifying what evidence does and does not say
• using evidence to draw my own conclusions
• knowing the difference between observation and inference
• understanding that there are different kinds of models and that the model is not the real thing
Build a Community of Scientists
• seeing myself as a scientist
• experiencing science as part of my daily life
• sharing and discussing findings with my classmates

The Triad Science Goals and their supporting strategies were articulated to aid teachers and scientists in engaging girls in science as scientists practice it in laboratories every day. For our community of teachers and scientists, it was an ongoing effort to see beyond the linear portrayal of science in textbooks and recapture the essence of science—wonder, questions, discovery, exploration, all leading to more questions. The Science Goals grounded our community in the habits of mind and activities that scientists engage in and focused our work on constructing opportunities for students to develop their inner scientist and to experience the wonder of inquiry, the excitement of asking questions, and the thrill of discovery. As opposed to the Triad Student Goals and Teaching Goals, these Triad Science Goals have undergone more revisions, moving from a traditional reflection of the singular scientific method presented in textbooks to a more modern representation of the nature of science as dynamic, evolving, wonderful, messy, incremental, creative, and personal (Lederman 1992). The Triad Science Goals are inspired by the efforts represented in the American Association for the Advancement of Science's Project 2061 and the National Research Council's *National Science Education Standards*. They are, however, distinct from content and knowledge standards (AAAS 1993; NRC 1996). Rather, these goals are a framework for developing a scientific culture in a classroom setting that can support rigorous, authentic learning of the scientific ideas articulated in local, state, and national science standards. Although the Triad Science Goals could have been derived in a variety of contexts, the gender-equity focus in our community is a driving force in the structure of this goal set and results in an emphasis on cultivating wonder, on the importance of community, and on the role of personal agency in science— for instance, thinking about what I already know and wanting to find out more. Throughout this chapter, you will find more information on each of the Triad Science Goals—Wonder about the Natural World; Do Science to Learn Science; Think Critically, Logically, and Skeptically; Use Evidence to Predict, Explain, and Model; and Build a Community of Scientists—highlighting the importance of these science skills and behaviors and how our own community has struggled to realize these in ourselves and our students.

SECTION II: Exploring the Triad Framework

Figure 4B

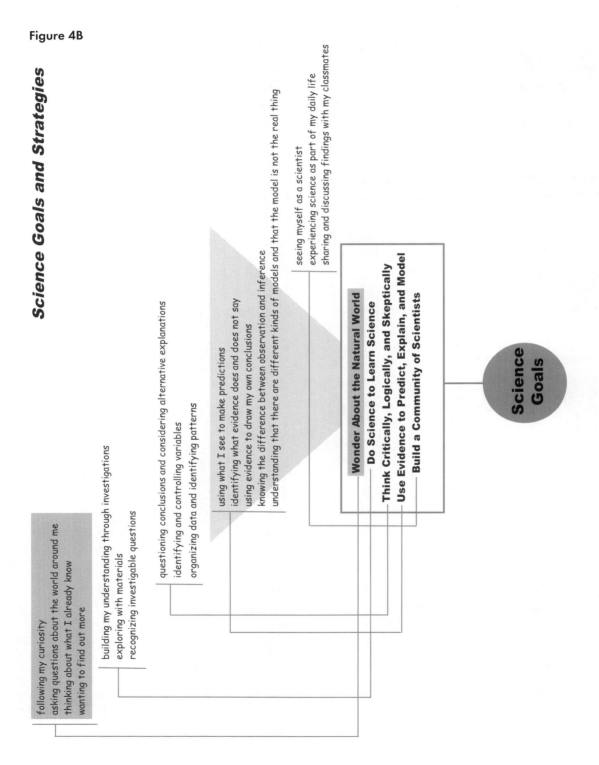

CHAPTER 4: Science Goals

Essay: Wonder About the Natural World

Although science is sometimes portrayed as a purely objective endeavor in which the scientist sets aside emotion and operates in a world of reason, the ongoing pursuit of scientific understanding is largely driven by wonder. If you don't experience wonder when you encounter a new phenomenon, true science, joyful science, passion-driven science cannot happen. It is nigh impossible to feel the drive to find out more in any subject, including science, without experiencing a sense of wonder, curiosity, and intrigue. Certainly science, like all disciplines (they wouldn't be called disciplines if they didn't require some work), can be tedious, frustrating, and difficult. Wonder about organisms, galaxies, or building machines drives creativity and new discoveries in biology, astronomy, and engineering every day. It's the wonder and the drive to find out more that make the details and hard work of science worthwhile to the scientist and support him or her in persevering through difficulty, frustration, and setbacks.

Often times, our own wonder is quieted by ourselves or by others—because of time, resources, or the perception that the joy of wonder is somehow antithetical to the serious intellectual work of science. In classrooms, wonder on the part of teachers and students may be lost in science due to the demands of focusing on standards, curricular constraints, and the challenges of large and diverse classes. So often in science classrooms, students do experiments or investigations completely in the absence of wonder. Few classroom science experiences are driven by student interest, questions, or wonder, so it is no surprise that students find it difficult to muster wonder. For girls in particular, engaging in wonder has the potential to connect science with the personal and the affective and to challenge them to reconsider the common stereotype of science as impersonal and objective. Our own questions and the things that we wonder about are indeed what we are most likely to engage with and learn from, and some would argue that wonder and having wonderful ideas is "the essence of intellectual development" (Duckworth 1996). In her essay on this topic, "The Having of Wonderful Ideas," education researcher Eleanor Duckworth recounts a study examining what young students would do when put in a roomful of materials with no teacher. The results suggested that those students whose classroom experiences had included an experimental program of open-ended, student-driven exploration had a greater diversity of ideas and pursued them to a greater depth than those students in the comparison classrooms (Duckworth 1996). Although it is unreasonable to expect that all students will find wonder in everything, we can strive to develop students who are able to find aspects of wonder in things all around them. In science,

SECTION II: Exploring the Triad Framework

then, wonder is a skill, just like observation. As such, wonder must be cultivated and practiced by students, just like any other skill, and explicitly taught as an integral part of the process of science.

All of us have things that have filled us with wonder at one time or another—perhaps it's thinking about how a single cell can ultimately become a human infant or that plants can make their own food from air, water, and sunlight or the beauty of a flower, skull, or bird's nest. So, how do we encourage and entrain our students to wonder? To encourage students' wonder about the natural world, we can share our excitement and enthusiasm about the natural world with them. We can have a classroom filled with materials ripe for exploration and provide time for students to explore materials (with all their senses when appropriate). We can allow time for questions, like bubbles, to percolate to the surface and keep an open mind about what arises. Often the wonderments that students bring may not be the same we adults might have, but we can recognize the importance of their questions to them, coach them in owning their ideas, and encourage them in pursuing them further. And we can engage students in reflecting on what they already know and wonder about within a topic and, perhaps most important, encourage students to own, pursue, and feel the drive to find out more about their own wonderments and questions.

Not What We Had Planned

One of the most rewarding lessons I got through the Triad experience was one I did not expect to learn at all. I decided to participate in Triad because I wanted to learn more about teaching and to see if I liked it enough to pursue it later on. I did not expect to learn anything new about being a scientist. However, it was often hard to keep kids' interest when teaching an activity, and I realized that I had to get to the root of why something was cool if I wanted them to get it. For example, when we tried to teach the kids about learning by classically conditioning some honeybees, they were much more excited about playing with the bees than about training the bees to do anything. They wanted to know how the proboscis worked, where the honey came out of the bee, and what the bees ate—good questions, but not what we had planned. However, I was reminded that science really is just about observing the world, asking interesting questions, and playing with them until you get a satisfying answer. As a research scientist, I easily get wrapped up in the details of my work—which statistical analysis to use, how to get a technique to work, or what journal to try to publish a paper in. The kids at Triad reminded me that I actually have an awesome job. I get to play in a lab all day in order to answer a cool question. I came out of Triad having a better appreciation not only of teaching, but of science as well.

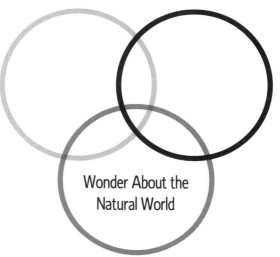

Wonder About the Natural World

Reflection Questions

- What did this scientist learn about her own profession from teaching science to young students?
- This scientist writes that the students during the lesson came up with "good questions, but not what we had planned." How would you respond in this situation?
- In what ways do you cultivate curiosity about the natural world in your own students? How can we structure lessons to encourage students' curiosity and use it to drive science lessons?

SECTION II: Exploring the Triad Framework

- Many approaches to planning science lessons promote *engagement* as an essential prerequisite for science learning. How do you begin your science lessons? In what ways do the beginning of your science lessons encourage students to wonder and be curious?

Links

- Student Goals: The Stopwatch as a Tool
- Science Goals: To Simply Marvel
- Science Goals: When Science First Was Really New
- Teaching Goals: See What Happens
- Teaching Goals: Theory Is Easy and Practice Is Difficult
- Teaching Goals: Personal Development

CHAPTER 4: Science Goals

To Simply Marvel

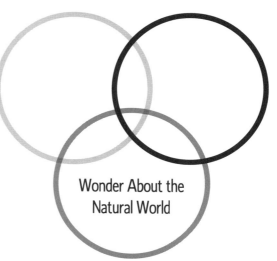

Wonder About the Natural World

I often feel while I am teaching class that students don't have adequate time to simply marvel and be in awe of the beauty and diversity of nature. Marveling and being in awe is what prompted me to study biology, and therefore I thought I'd like to try to inspire a little wonderment in the girls of the club. I created a lesson by collecting seeds from all around that I thought had some interesting features—edible, indigestible, huge, tiny, interesting seed casings, and interesting ways of being transported, to name a few. I placed the seeds on the table and we had a discussion of seeds—what they are and their purpose—as the girls ate the seeds.

Reflection Questions

- Imagine the lesson about seeds described above. Do you predict that the students were marveling and in awe of the seeds? Why or why not?
- Describe an example of when you have seen students in your own classroom marvel and wonder at the natural world. What prompted the wonder? What did it look like?
- Is it possible for wonder to occur when the topic for learning must be determined by the teacher? Why or why not?
- Where are the places in your curriculum where student-driven learning and time for wonder could occur? How would you approach integrating more wonder into your science lessons?

Links

Student Goals: No Longer the Same

Science Goals: Not What We Had Planned

SECTION II: Exploring the Triad Framework

- Teaching Goals: Talking in Questions
- Teaching Goals: Resurrecting Socrates

CHAPTER 4: Science Goals

The Balloon Droops

This vignette is from notes written by an observer at a mixed-sex, after-school science club for sixth, seventh, and eighth graders in which the students were exploring the results of a baking-soda-and-vinegar reaction.

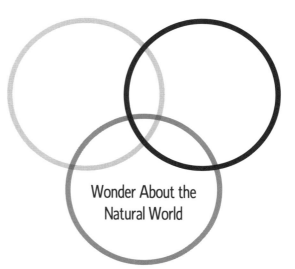

The club members were putting four Alka-Seltzer tablets into a balloon, attaching the balloon to the top of a soda bottle, and tipping the tablets into the vinegar waiting in the bottle. This was a bit of a difficult task. The tablets were big and had to be broken up, usually resulting in a powdery mess. The students, however, were all excited to be doing it, as balloons around the room began to inflate. After the first time through the activity, they started experimenting with variations—mostly more of each ingredient, but some tried other things. One of the sixth-grade girls suspected that the gas produced might be heavier than air because when one held the bottle horizontally, the balloon drooped. She brought this idea to one of the club-sponsor scientists and then decided to try and compare the balloon on the bottle with a regular balloon. She blew up a balloon to a size equivalent to the experimental balloon, tied it up, and then tied up the experimental balloon. When she dropped both balloons, the experimental one hit the ground a little bit faster. She concluded that the gas from the Alka-Seltzer-and-vinegar experiment must be a heavier gas.

Reflection Questions

- Based on the description above, how would you describe the classroom culture in this science club?
- How can student curiosity and exploration be encouraged, especially among students who have been acculturated to following directions and arriving at predetermined answers?
- Science as done by scientists is messy and nonlinear, though this is not typically how it is portrayed in textbooks. In what ways is the process of science portrayed in your curriculum and your own science lessons?

Girls in Science

SECTION II: Exploring the Triad Framework

- To what extent do students in your classroom have the opportunity to wonder and explore their own ideas and be doing different things within the context of the same lesson?

Links

 Student Goals: Not Having Step-by-Step Instructions

 Science Goals: There Is No Road Map

 Teaching Goals: Talking in Questions

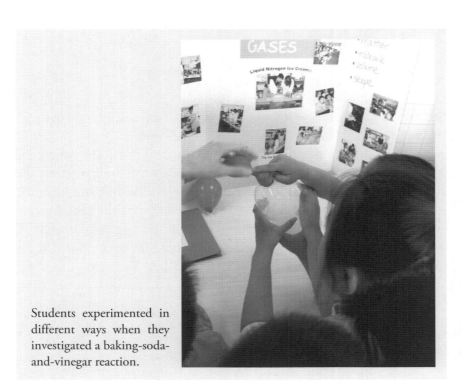

Students experimented in different ways when they investigated a baking-soda-and-vinegar reaction.

CHAPTER 4: Science Goals

Above a Whisper

For our final lesson, we took the club on a sailing trip around the bay for an hour. It was cold and foggy, and I was worried about some of the students who had brought only a thin sweater despite our warnings about the weather. The kids, however, did not seem to mind the cold, because they were so excited. One of the girls told me that she could not sleep the night before because she was up thinking about the trip. I watched them crowd around as we lowered a net into the bay to collect plankton and look in wonder when a seal surfaced near the boat. One girl, who had never spoken above a whisper in school, jumped up and down and cried out to people on the shore as she waved at them. It made me realize how important it is for kids to experience learning outside a typical classroom setting, to talk to real live scientists, to experiment with real animals, chemicals, and electricity, rather than to just read about them in books.

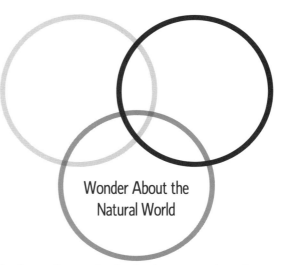

Wonder About the Natural World

Reflection Questions

- What did the scientist learn about the students from seeing them in this different setting?
- What types of science experiences have you seen spark the most wonder in your students? Why?
- What is the role of novelty, new learning situations, and new environments in increasing students' wonder in science? How do you integrate the new and the novel into existing curricula?
- How do you teach students to cultivate wonder in themselves, even in the context of an everyday classroom? In the context of reading about science?

Links

- Student Goals: By the End of the School Year
- Student Goals: The Real Microscope

SECTION II: Exploring the Triad Framework

- Science Goals: The Real Thing
- Teaching Goals: The Quieter Girls
- Teaching Goals: Science Is Not a Priority for These Students
- Teaching Goals: By Scoring When a Girl Participated

Student Work

1. Describe **OR** draw your favorite Triad activity. Include as many details as possible. (If you draw, please label your picture with words.)

We went scilling on the Ruby. It had sails, and fun nooks and crannies

2. Why was this your favorite activity?

Because I love the ocean and the thrills of it. I love sailing because I went last year and I got over the fear of falling over. Because of this expeirience, I appreciate ships more & this was really, really fun. There was a bit of excitement and thrill in it.

CHAPTER 4: Science Goals

Nothing to Do With the Club

My teacher partner and I spent several hours on the telephone refining our activity plans. Sometimes we talked about science that had nothing to do with the club. Here's an example of one e-mail discussion between my teacher partner and me.

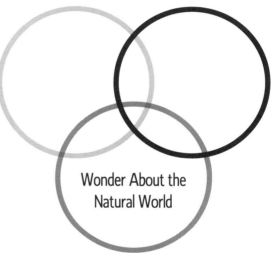
Wonder About the Natural World

> Hi Frances, I had a wonderful time yesterday. I was extremely impressed with the girls' curiosity and creativity. I am still feeling an incredible high over the whole thing. Thank you! I thought more about your question about why red light plus green light makes yellow light. I haven't checked in a textbook, but I think I know what is happening. I think it has to do with the interference between the two colored beams producing a new beam of another wavelength. Red light has the longest wavelength in the visible spectrum, and violet has the shortest. Green has something in between. When green and red light "mix," the interference of the waves changes the wavelength of the light to something in between red and green. That's why the color you see is yellow. Does this agree with other observations you have made? Trisha

That it was a genuine scientific discussion was evidenced by her disagreement with me on this matter in her return e-mail.

> Trisha, you have a reasonable explanation. I am wondering about how the human eye works. I think (need to look it up) that we only detect red-green-blue. I am still thinking that it is mostly in the perception. Frances

Reflection Questions

- Even the most experienced science teachers and scientists understand very little about how the natural world works. To what extent do you take time yourself to wonder about the natural world? Why or why not?

SECTION II: Exploring the Triad Framework

- Is it possible to cultivate wonder in science in our students without first cultivating wonder in ourselves? Why or why not?
- For you, what is the role of working with others in inspiring and sustaining wonder?
- What strategies have you used to teach your students to wonder? to acknowledge their wonder? To develop questions from their wonder? To pursue further the questions that arise from their wonder?

Links

○ Student Goals: On a More Personal Level

◐ Science Goals: Walking Encyclopedia

● Teaching Goals: Resurrecting Socrates

SECTION II: Exploring the Triad Framework

Figure 4C

Science Goals and Strategies

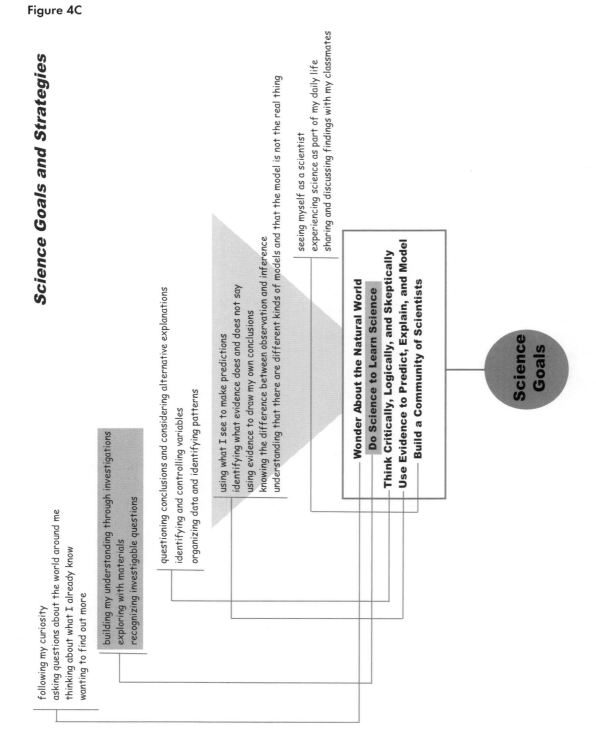

CHAPTER 4: Science Goals

Essay: Do Science to Learn Science

If we want students to see science as interesting, exciting, and relevant to their own lives, we need to help them experience the thrill of discovery, the confusion of contradictory data, and the wonder of new insights. Otherwise put, knowing the history of human uses of pottery does not a potter make. One must learn to throw a pot. In this same spirit, knowing the historical discoveries of science does not a scientist make. One must learn to discover ideas. At its core, doing science begins with tinkering and exploring, playing and wondering, following either one's curiosity or one's conundrums. It goes from there to identifying questions that can be answered through scientific investigations and then designing appropriate investigations. Different kinds of investigations yield different kinds of evidence that needs to be mulled and worked through, and new questions and hypotheses evolve with these new findings.

Teachers have long known that students are more engaged in science when they don't learn it solely from a book. By actually testing the electrical conductivity of different substances, students are involved in a different way than they are by reading about good and poor conductors in a textbook. But it is not only motivation that is enhanced when students try to find out about the world around them by interacting directly with it. They also learn more science. This idea is central to the *National Science Education Standards* (NRC 1996). At the core of the Standards is the concept of inquiry, namely that students should have more opportunities to engage in science in ways that reflect what scientists do in laboratories. Girls, however, often have fewer opportunities in classrooms to engage in hands-on science to learn science. Research suggests that boys more often than girls have their hands on science materials and are the drivers of hands-on experimentation in classrooms (AAUW 1998), a finding re-confirmed by participants in our community many times in their own classrooms. In addition, one is hard-pressed to find female scientists who do not have at least one story in which they deferred to male peers during classroom investigations during the secondary, college, and even graduate school experiences.

When envisioning what Doing Science to Learn Science may look like for students, hands-on investigations may be what first comes to mind. There are many times when students can engage in direct, scientific inquiry—investigations that originate with their own confusion, with a wonder-induced question, or with a discrepancy between how they view the world and what their textbook says. In fact, they sometimes can find things out that no one else knows. For example, they can investigate

SECTION II: Exploring the Triad Framework

the variety of species of plants and animals that live in a nearby vacant lot or in a grassy area of the school yard. This kind of investigation helps them understand how scientists work: how they decide what kinds of data to collect (can they sample just a few parts of the lot? what constitutes an adequate sample?); how they count and collect material (what tools do they need? when do they look? how far beneath the surface should they dig?); how they can be as accurate as possible (how do they check their observations?); and how they decide about communicating their findings to someone else. Wrestling with these kinds of tasks helps students understand what scientists actually do—and it also enriches the concepts that they learn.

At its core, achieving this science goal is about structuring classroom learning in such a way that students have opportunities to design their own investigations to answer their own questions. Our challenge as educators is to decide how much time to devote to actual investigations by students—whether original or not—and how much to read about how scientists have conducted inquiries that have led to the ideas that we hold today. For such decisions, local and practical considerations are paramount. Is material available for hands-on investigations? How much time must be taken to prepare the students for field or laboratory work? Teachers can set the stage for the investigations, provide the topic(s) and materials for the investigations, and help the students think through their investigations while allowing students the freedom to pursue their own questions. The extent to which classroom science should be teacher-driven or student-driven can vary widely (NRC 2000). However, it is through identifying questions that they care about answering, developing ways to gather evidence, trying to make sense of their observations, and sharing a new understanding with others that students begin to function as scientists. In this way, this science goal is fundamentally about aspiring to have school science bear a greater resemblance to science as practiced by scientists in laboratories every day.

CHAPTER 4: Science Goals

Putting Sugar in Water

At a recent meeting, a colleague talked about the fact that science transcended laboratory walls and extended into many places like the kitchen, where every time you dissolve sugar or salt in water you "do science." I felt that some people thought this thinking was progressive—outside the box. The idea that some people might have thought this example progressive—whether they did or not—bothered me.

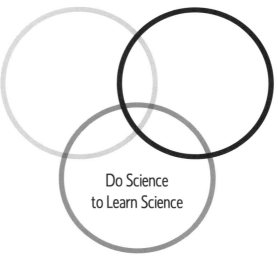

I've always hated that example of "doing science." And I thought it bothered me because kids already tend to think that chemistry, and only chemistry, is science. I have observed this misconception via student complaints that we hardly ever get to "do real science," accompanied by longing gestures toward the large glass flasks and test tubes on top of the two tall cabinets in my classroom.

Most recently, a few second graders doing pill-bug investigations came over to ask my permission to "do an experiment." They were looking over at two boys who were mixing alum, water, toothpaste, Kool-Aid and various other things in a vial (in order to try and figure out if alum separates all things into different layers or just the soil we had first tried it with). Their eyes—longing and gazing toward these two boys and the vials they were shaking furiously—told me what type of experiment they wanted to do. I explained, "But you're already doing an experiment." They looked very confused. "You're doing a biology experiment with pill bugs," I elaborated. They still looked puzzled. "But, we want to do what they're doing," they pointed. "Ohhhhh, you want to do a chemistry experiment," I said. They smiled and nodded, hoping this meant they would get to go make a mess. "Do you know what chemistry is," I wondered aloud. *Mixing* as well as the wonderfully mysterious and oh-so-exciting-word *chemicals* came up. I wish I remembered their precise definitions.

So, I thought this example of dissolving bothered me, because if we use it with kids, it serves to strengthen this misconception that chemistry—and more specifically creating chemical reactions—is the sole branch of science. But now I think it bothers me much more, not because it perpetuates that myth, but because I don't even think

SECTION II: Exploring the Triad Framework

dissolving sugar in water is necessarily doing science. If you're putting sugar in water to sweeten your tea, you're not doing science anymore than when you are breathing, walking, or being. I think that the only time putting sugar in water is doing science is if you're putting it in to observe and see what happens to the sugar, or the water, or both. Doesn't doing science depend on the intent of what you're doing?

Reflection Questions

- "Doesn't doing science depend on the intent of what you're doing?" To what extent do you agree with this teacher's definition of "doing science"? To what extent do you disagree? Why?
- What ideas about doing science do your students bring to the classroom? How are your students' ideas similar to or different from the students described above?
- What is an example of a science activity that students experience in your classroom that actively challenges their pre-conceptions of what "doing science" means? What is an example of an activity that may perpetuate student misconceptions about science?
- What are specific teaching strategies that we can use to move students away from a concept of science as only mixing and mess? Consider how engaging students in making predictions, asking questions, sharing evidence, and drawing conclusions could be influential.

Links

Student Goals: No Longer the Same

Science Goals: I Learned How a Lava Lamp Works

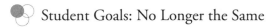
Teaching Goals: Does This Bridge Look Better Than It Did Last Time?

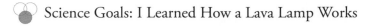
Teaching Goals: Accepting Stereotypes

I Learned How a Lava Lamp Works

One example of what I have learned is to keep the students aware of your goals for the lesson and the students. I had always thought it is best to let the students to figure out for themselves what the point of a lesson or activity is. I see now that making the students think about what you have designed the activity to teach provides a context into which they can put their new discoveries and knowledge. If the activity is designed well, they will have plenty to discover and think about and shouldn't be bogged own with won-

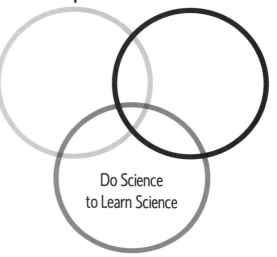

dering, "What is the point?" One science club activity my group led was lava lamps. By being explicit about what they could learn from making and playing with a lava lamp—what does density mean and how do liquids of different densities interact—(hopefully) we helped them come away from the activity saying "I learned how a lava lamp works" rather than just "Today we made lava lamps."

Reflection Questions

- How would you structure a lesson on lava lamps so that at the end students could say, "I learned how a lava lamp works," rather than "Today we made lava lamps?"
- Describe a science teaching experience where you have had students ask, "What is the point?" or "Why are we learning this?" How did you respond?
- How does this scientist see how sharing the goals can be a part of the learning process for students?
- Do you explicitly share the goals of your science lessons with your students? Why or why not? If so, how do you integrate the goals of the lesson into the learning experience of students?

SECTION II: Exploring the Triad Framework

Links

- Student Goals: Not Having Step-by-Step Instructions
- Science Goals: Putting Sugar in Water
- Teaching Goals: The More We Expected
- Teaching Goals: Talking in Questions
- Teaching Goals: At First I Was Hesitant

The goal was for students to learn how a lava lamp works, not just how to make one.

CHAPTER 4: Science Goals

Nobody Knows What's Inside

These vignettes are from an observer's group interviews with sixth-, seventh-, and eighth-grade students in an all-girls club. They are responding to a question about which activities they have enjoyed doing most.

Sixth and Seventh Graders:

Jayalan: The mystery box—that was kind of messy, but that was fun.

Interviewer: What made the mystery box fun?

Jayalan: Well, at first we were really anxious to know what the scientists had done to their mystery box. So then, we just decided to make our … well, they told us to make our own, and it was fun to do that. It's like nobody knows what's inside. And it's so mysterious.

Erin: Yeah, that's what I picked it as my favorite, too, because … all of us tried different patterns to get the colors, and nobody knows the way we did it. Everyone was, like, how did they do it? We tried to figure it out. That's why it was called the mystery box.

Rachel: We could use other materials, and we had to figure out how to change the colors. And it was fun because we got to see the inside of the mystery box when we were done.

Interviewer: Was that activity like what scientists do, do you think?

Erin: Actually, I think that ours is more fun. Theirs … you have to be serious—you can't really laugh about things. But I could laugh, talk a little bit, have fun. For them, one small mistake and everything's done.

Eighth Graders:

Doreesh: I like the magic boxes, even though we didn't finish—my group didn't finish. That was really fun, doing it as a team, trying to find out what you could do. I really liked it.

Sing: In middle school … we don't really get to do those things in school, like making stuff a lot. I mean, we just study, and do science activities. But now, it's like you get into it more than you usually do in science.

SECTION II: Exploring the Triad Framework

Rose: Yeah, I liked the mystery box, too, cause it's more like figuring out stuff, and you don't really have directions.

Doreesh: It's like starting from zero to make something.

Reflection Questions

- Given what the students describe, how does the mystery-box investigation represent the doing of science? How does it not? Would you let students look inside the mystery box at the end of the lesson? Why or why not?
- Rose said, "You don't really have directions." Doreesh described the mystery box as "starting from zero." Describe a science lesson that you teach that is most similar to the mystery box, namely a lesson that is more inquiry based and in which students drive more of the decisions about the investigation than the teacher does.
- What are strategies that you have used to transform science demonstrations and cookbook laboratory activities into more inquiry-based, open-ended science investigations?
- How do you think the nature of the activities students experience in their school science affects their perceptions of the profession of science and the doing of science by real scientists?
- Doreesh said, "I like the magic boxes even though we didn't finish." What about this activity might have been satisfying for Doreesh even though her group was unsuccessful in finishing the task?

Links

- Student Goals: Fun and Frustrating
- Student Goals: To Build and Rebuild
- Student Goals: Not Having Step-by-Step Instructions
- Teaching Goals: Theory Is Easy, Practice Is Difficult

Students liked "figuring out stuff" without directions when the class made mystery boxes.

CHAPTER 4: Science Goals

The Real Thing

Another professional development activity that consolidated many of my thoughts about teaching was the mystery-fruit activity in which it was clearly illustrated that hands-on experience with an object far outweighs any photograph, text passage, or diagram. You could also see this in the difference between the students' interest in our two cow-eye dissection lessons, the first of which had eye models and diagrams while the second was a dissection of a preserved eye. The students were much less interested in the models than in the real thing. But the models helped start the learning process so that they were familiar with how an eye works, its features and organization, so that they could immediately jump into the dissection. Simply knowing a fact does not mean that you can apply it, and, clearly, no learning is as engaging or memorable as hands-on experience.

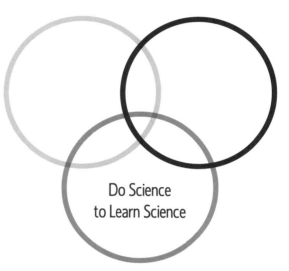

Do Science to Learn Science

Reflection Questions

- What did this scientist discover about the role of the real thing in science teaching?
- What is your earliest memory of learning science? Why do you think it is so memorable for you?
- Drawing on your own experiences, compare and contrast what you have observed about student learning that happens when students are exploring with concrete materials versus reading about science in a textbook.
- Describe a science lesson that you have transformed from being primarily book learning to being more about exploring with materials. What did you notice about student learning using these different teaching approaches with the same content material?
- In your own teaching, when do you use models, diagrams, textbooks, and other science references previously generated by others? Is it primarily at the beginning, in the middle, or at the end of a science lesson? Why?
- How does the timing of using more abstract information affect students' interest in or willingness to explore with materials and build their own ideas?

SECTION II: Exploring the Triad Framework

Links

- Student Goals: No Longer the Same
- Student Goals: After the Initial Eeewwww
- Student Goals: The Real Microscope
- Student Goals: Safety Was a Concern
- Teaching Goals: I Could See How Much I Learned
- Teaching Goals: A Different Role

CHAPTER 4: Science Goals

A Daunting Task

Although I can clearly articulate my hopes for the achievements of my science club, I cannot say with certainty that I can articulate my overall educational philosophy. For the science club, I think that science process should be the sole focus and goal. Overall, however, content is important. Science process cannot stand alone. Perhaps it is possible to emphasize science process while integrating science content. For example, analyzing data, coming to conclusions, and making predictions—all introduce science content into science process. But executing the science process takes time and resources. To teach all science concepts using a hands-on, science process–based approach is a daunting task. Perhaps more traditional ways of teaching science content are the only efficient way to convey this information.

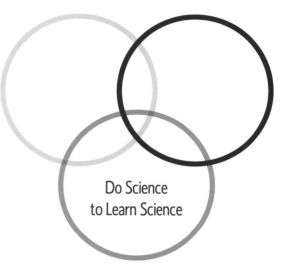

Reflection Questions

- How is this scientist struggling with the goals of her science teaching? In what ways is this scientist's struggle familiar to you? In what ways is it foreign to you?
- Compare and contrast what the following terms mean to you: hands-on science, inquiry-based science, minds-on science, lecture-based science. How would you characterize your own philosophy of science teaching using these (or other) terms?
- For which science concepts have you found it most critical that students have opportunities to learn in a hands-on, materials-intensive way?
- What are some ways you could engage students in the doing of science in the complete absence of hands-on materials?
- Do you see a dichotomy between teaching science process and science content? Why or why not?
- In the vignette above, the scientist states, "Perhaps more traditional ways of teaching science content are the only efficient way to convey this information." What is the relationship, if any, between student learning and efficiently conveying information?

SECTION II: Exploring the Triad Framework

Links

- Student Goals: No Longer the Same
- Science Goals: There Is No Road Map
- Teaching Goals: Theory Is Easy, Practice Is Difficult
- Teaching Goals: Back in the Classroom

SECTION II: Exploring the Triad Framework

Figure 4D

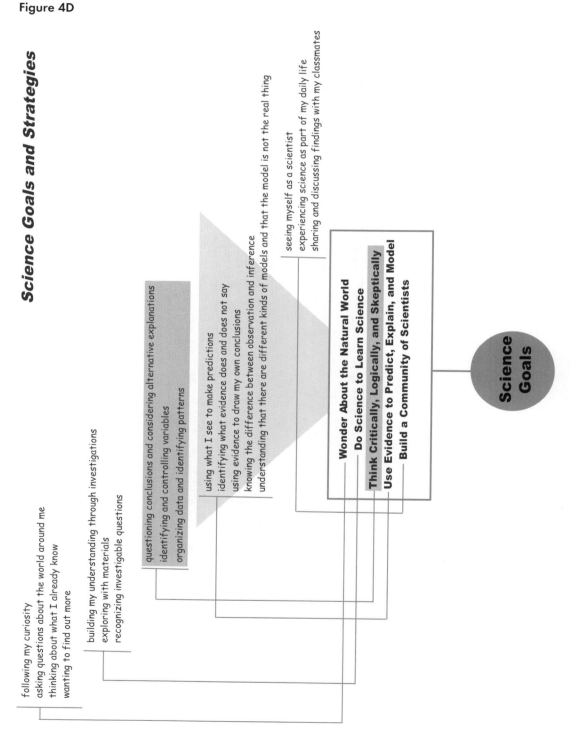

CHAPTER 4: Science Goals

Essay: Think Critically, Logically, and Skeptically

Life in the 21st century is fairly characterized as a constant bombardment of information from other people, television, the internet, phone calls, e-mails, faxes, newspapers, magazines, and more rarely books, sometimes originating from unknown sources with unknown agendas. When probed on how they've come to know something, many people will give answers like: "Well, I read it in a book." "It said so on the internet." "They say that's how it works." This information bombardment affects young and old alike and reaches into almost every corner of the earth where humans reside. It is an essential life skill to be able to evaluate information—to think critically about claims made and conclusions drawn; to analyze information logically by detecting alternative explanations and anticipating logical implications; and to consider information skeptically and evaluate it based on its supporting evidence, its source, and one's prior knowledge.

Though useful in many arenas of knowledge, thinking critically, logically, and skeptically is at the core of scientific habits of mind. Scientists must always be analyzing, evaluating, and judging the evidence at hand, on the critical lookout for any other explanations that could provide a different explanation. They are engaged in logically reasoning through data and observations from multiple investigations and deducing general principles. They are skeptical in accepting conclusions, often making predictions that would have to hold true for them to accept a claim made by other scientists. Thinking critically, logically, and skeptically in science is a driver of new ideas, producing previously unthought-of explanations and paradigm shifts in the way we understand the world. Throughout the history of science, there are tales of creative thinkers proposing new ideas of how the world works. The role of scientist is not an easy one, often conjuring up fierce criticism and initial ridicule from other scientists in proportion to the novelty of an idea. Novel ideas arising from scientific thinking can themselves become prey to over-enthusiastic criticism and skepticism. To understand the natural world, scientists and all of us must be able to be simultaneously awe-filled and cultivate these scientific habits of minds—thinking critically, logically, and skeptically—that check and balance our evolving understanding of how the world works.

Due to a variety of practical and institutional constraints, students in the U.S. educational system may often be more focused on remembering the information they receive at school and not inclined to question or doubt it. Practically, students in the

SECTION II: Exploring the Triad Framework

United States are presented with more information in their science education than most other students around the world, covering more topics in math and science before high school than 75 % of countries studied in the Third International Math and Science Study (Mullis et al. 2004). The current priority of quantity over depth of scientific information learned in our country puts both teachers and students in the challenging position of struggling to make the time to be critical, logical, and skeptical about the ideas in science they are studying. Students can be so focused on what answer the adult in the room wants from them that they abandon both their own wonder and their own power to accept or reject any idea—in science or any other subject—being presented to them as fact. The bright and eager curiosity of the preschool years fades in the face of tomelike textbooks and multiple-choice tests grounded in right answers. Developing a balance between sharing with young students what is known about how the world works and cultivating thinkers who are critical, logical, and skeptical is a classroom ideal that can be difficult to achieve. The science classroom, though, is fertile ground to revitalize both students' sense of wonder and their sense of power and ownership of knowledge by encouraging them to hone independently their skills of thinking critically, logically, and skeptically about incoming information and explanations.

So, what does teaching students to be critical and skeptical look like? Some of the strategies that have emerged from the Triad community include:

- Encourage students to see knowledge as ever evolving. Help students understand that the scientific body of knowledge is always a work in progress. Although there is much that scientists agree on, there is also much disagreement in the scientific community as scientists question and think about the evidence. It is not unusual for ideas that have been accepted for many years to change in light of new evidence. For example, it was long held that nerve cells do not reproduce themselves in adult humans; with new techniques and some critical, logical, skeptical thinking, experiments were performed that showed that a small number of nerve cells do divide in adult humans.
- Practice developing alternative explanations. The concept of controls and variables in science classrooms can be an ongoing challenge for teachers and students, and yet controls and a consideration of the variables at play in an investigation are just a formalization of ruling out alternative explanations and applying logic in science. If the plants in the experiment had different amounts of light and different amounts of soil, then we can't conclude that their ability to photosynthesize is dependent only on light. We need to control the amount

of soil so that it is not a changing variable. That way, if we see differences in photosynthesis among plants in different amounts of light, we can rule out soil as an alternative explanation. Scaffolding the links between these connections for students as they are learning to think logically is helpful.

- Cultivate the habit of questioning in students. Questions are paramount in science, and the willingness to question is a tool of the trade. Yet, often students are more practiced in answering questions than asking them. Reflective journals can prompt students to record and value their questions. A requirement of at least three questions (and from new voices each time) during any classroom presentation or discussion can build a culture of questioning. Practice, practice, practice can lead to the questioning student and the development of critical, logical, and skeptical thinking.
- Keep a focus on evidence. As discussed in the essays "Defending a Position," p. 91, and "Use Evidence to Predict, Explain, and Model," p. 147, evidence is the currency of science. If we expect students to think critically, they must be given access to primary evidence, results from investigations of their own, their peers, and/or from practicing scientists who may have made the discoveries they are learning about. In addition, they must be given the license to be skeptical, brutally skeptical, as long as they are earnestly evaluating the ideas and evidence at hand. To give students these opportunities to be critical, logical, and skeptical, we as teachers must be brave and know that questions will be asked that we cannot answer, that alternative explanations may be proffered that we cannot rule out, and that at the end of a unit many unanswered questions and pending ideas will remain, just as in any scientific endeavor.

SECTION II: Exploring the Triad Framework

Answers Are Not the Goals

Triad taught me that information and answers are not the goals in and of themselves. Questioning, thinking, speculating are more important. I learned this from modeled behavior by the Triad staff and from watching my partner teachers lead wrap-up discussions during science clubs, in which they encouraged the students to ask questions about what they had seen and to speculate about what might be going on. Largely, the teachers refrained from comment. They simply facilitated the students' brainstorming. I think that this helped the students to realize that adults don't inherently have all of the answers—especially since we were not experts in most of the subjects we studied—and to learn that science works by asking questions and making models.

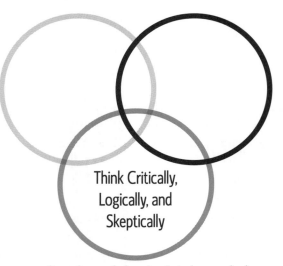

Reflection Questions

- What common beliefs about science contribute to a valuing of "information and answers" over "questioning, thinking, and speculating" in many science learning environments?
- Describe a time in which you were given the answer after you had worked to solve a problem. How did this information affect your feelings about your work?
- When is the introduction of scientific information and answers important? When might it be counterproductive to give students known scientific information? What impact might the timing of this information have on students' perceptions of their own abilities to think critically?
- If our goal is to get students to think critically, logically, and skeptically, how should we respond to student questions and facilitate discussions? What were the teacher partners of this scientist doing?

CHAPTER 4: Science Goals

Links

- Student Goals: I Assumed That Our Girls Would Feel Comfortable
- Science Goals: The Most Difficult
- Science Goals: Scientifically Dissatisfied
- Teaching Goals: Talking in Questions
- Teaching Goals: Resurrecting Socrates

Walking Encyclopedia

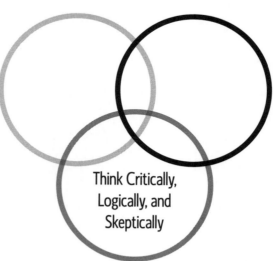

I continue to be struck by how accepting most girls—and teachers for that matter—are of any information that comes from a scientist. People often share with me things they've heard about in science, often starting with the phrase: "Well, they say that ..." *They* is apparently a term used to describe anyone who claims to be a scientist. Scientists are automatically assumed to be all knowing, and therefore anything they say must be true. As a scientist, I have tried to demystify science and counter the perception that only the brilliant do science. I have tried to encourage everyone I meet to question and be skeptical of all the information they hear, even if it comes from scientists.

My work certainly involves asking questions and doing experiments, but the real work of science comes when you're trying to analyze evidence and make conclusions. My adviser likes to say that if you give a piece of data to 20 scientists, you'll get 20 different interpretations and explanations. The ideas would certainly be related, but each would be filled with wonderful nuances and particular insights about what could explain what was going on in the experiment.

Among my Triad scientist colleagues, there were often conversations about how we sometimes felt we were expected to be walking encyclopedias, fountains of knowledge on all scientific subjects from all disciplines. I didn't want to disappoint kids and teachers by not answering their questions, but I also don't want to propagate the notion that scientists are walking encyclopedias.

Reflection Questions

- To what extent do you question scientific information you read in the news or hear about from others? How do you cultivate skepticism in yourself?
- A fellow science teacher at your school tells you that she just found out something interesting about a common candy. She heard that it causes mad cow disease because it is made in Great Britain and it contains gelatin. How would you

CHAPTER 4: Science Goals

respond to your friend? Do you believe that she had given you evidence that the candy causes mad cow disease? What other information might you need to find out to be able to think critically and skeptically about this claim?
- What do you think contributes to the common perception that scientists know the answer to any question, that their knowledge generalizes to areas outside their own fields of study, and that scientific knowledge is truth?
- What kinds of science classroom experiences might help counter these common perceptions? What science activities have you specifically used to cultivate skeptical thinking among your students?
- Skeptical thinking taken to an extreme can result in rejection of evidence and an unwillingness to consider new information. How do we foster openness to new ideas and novel findings while also encouraging skepticism in students?

Links

- Student Goals: Where to Draw the Line
- Science Goals: The Most Difficult
- Teaching Goals: Like Dad
- Teaching Goals: Answering Student Questions With Questions

SECTION II: Exploring the Triad Framework

To Trust in Their Own Logic

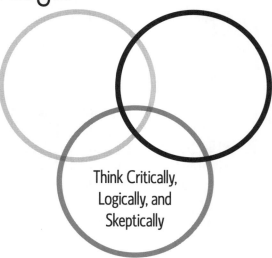

"What's the answer? What's the real answer?" Sometimes it seemed that no matter what we tried, our team couldn't avoid these kinds of questions from girls at the end of an open-ended, inquiry-based science activity. We seemed unable to get them to trust in their own logic and believe in themselves and their ability to reason through any evidence they collected or that we gave them. My teammates and I shared with them time and time again that science was about the collection of evidence and the accumulation of evidence and ideas over time. I told them that nobody was going to tell me the real answer to my thesis research question. I was, myself, going to bring to bear all the evidence I had collected about my question and draw my own conclusions in the context of what other scientists already knew. That was how real science worked.

The most striking example of my students' unwillingness to trust their own logic and reasoning came at the end of a discussion of a crime lab activity. My teammates and I had decided to construct the crime scene so that the evidence would point to at least two different suspects. There would be, in fact, no one right answer. The evidence would be degenerate, in the hopes of sparking debate among the girls. We had no idea what they would conclude, and, frankly, we didn't care, as long as they were using logic and evidence to make their claims and they were being critical and skeptical of one another's ideas. I predicted that they would conclude a team of people had committed the crime and left different types of evidence at the scene.

Although they did share their interpretations of the evidence and their conclusions, they were tentative in doing so. Their voices would rise in a questioning tone toward the end of sentences as they looked from one Triad adult to another for signs of whether they were on the right track. At the end of our discussion of the evidence, the group came to the consensus that three of the suspects were likely involved. I was feeling good. And then a pair of our students approached me and asked, "So who really did it?" I responded with, "Well, who tells the police who really did it, or the judges or the juries?" They looked at me disappointedly, and I realized that it was going to take a lot to get them to trust in themselves and their own conclusions.

CHAPTER 4: Science Goals

Reflection Questions

- What kinds of classroom experiences contribute to the common perception that every problem has a real answer that is already known? To what extent do you believe this to be true?
- What might students gain from wrestling with a problem for which they are never given the answer even if it is known? What effect might giving students the "real answer" have on their learning?
- To what extent do you think a focus on right answers might limit students' willingness to share their own ideas especially if they are struggling or confused?
- Think about the types of science investigations in your own curriculum. How many of these investigations have more than one possible result? How do your students respond to these more open-ended investigations?
- For so many science activities, students are asked to arrive at an already known scientific conclusion from evidence collected in a hands-on activity or demonstration. What strategies have you used to translate demonstrations or cookbook laboratory activities into more open-ended investigations that demand students think critically, logically, and skeptically?

Links

- Student Goals: We Have Reason to Believe
- Student Goals: Not Having Step-by-Step Instructions
- Science Goals: Scientifically Dissatisfied
- Teaching Goals: The More We Expected

There Is No Road Map

General and fundamental thinking can be learned by engaging in the process of science. Specifically, the process of science engages and calls on curiosity. Science process encourages, indeed even requires, critical thought. Curiosity and critical thinking are the basis of any academic work, be it liberal arts or science, and result in informed decision making on a day-to-day basis. For these reasons, I believe that giving girls an opportunity to conduct hands-on experimentation with the goal of exposing them to the process of science should be a main focus of science teaching. Science process is a way to train thinking and is a larger part of the fun that science has to offer.

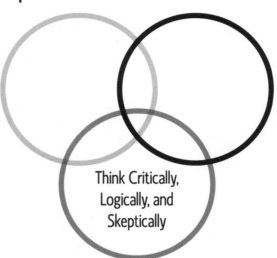

I wanted the girls to come away from each science club meeting having experienced some aspect of science process. Science process involves curiosity, interpreting data, drawing conclusions, and making predictions. Working as a scientist in a lab, I am involved in the process of science daily. In that lab there is no road map or exact protocol to take me through the process of science, unlike in so many lab exercises that I preformed in middle school, high school, and college. I wanted girls to experience science as it's done in the lab. I wanted them to have the opportunity to think about what to do in an experiment, not just follow a set of directions that someone else had described. I think experimentation is an important life skill that the girls will use in lots of ways besides science class.

Reflection Questions

- Describe a time from your own experiences when you were fully engaged in thinking critically about a problem. What did this experience demand from you and develop in you that following a detailed lab protocol would not?
- "Critical thinking … results in informed decision making on a day-to-day basis." How do you connect the skills of science thinking to everyday critical thinking for your students?

CHAPTER 4: Science Goals

- What kinds of scaffolding might be helpful to students who have never had experiences figuring out their own road maps to solve a scientific problem?
- What are some challenges that classroom teachers face in providing students with practice in thinking critically? What are some strategies for managing these challenges in the classroom?

Links

- Student Goals: Not Having Step-by-Step Instructions
- Student Goals: The UV Bulb Can Be Changed by the User
- Science Goals: A Daunting Task
- Science Goals: The Balloon Droops
- Teaching Goals: I Could See How Much I Learned
- Teaching Goals: Does This Bridge Look Better Than It Did Last Time?

SECTION II: Exploring the Triad Framework

Figure 4E

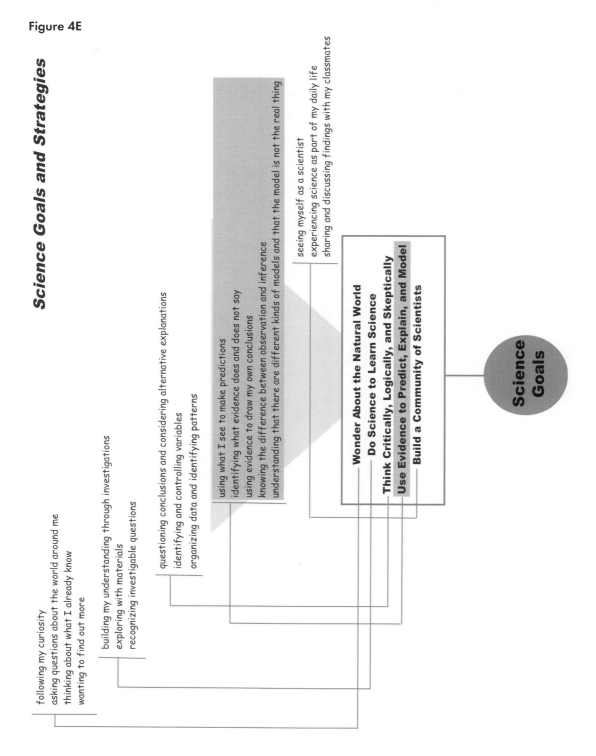

CHAPTER 4: Science Goals

Essay: Use Evidence to Predict, Explain, and Model

Scientific descriptions of the natural world are all grounded in evidence. The use of evidence is an essential element that distinguishes science from other ways of knowing. Scientists can speculate all they want about how the world might work, but how it actually works can be shown only through evidence—reproducible evidence. This evidence is systematically collected and can be done so through a variety of investigational approaches including observation studies, collection analysis, and experiments. Evidence can be descriptive or quantitative and often consists of noting particular patterns under defined conditions. Evidence can also come from credible published sources such as peer-reviewed journal articles, books, and databases. The value of evidence in the scientific world lies in its capacity to guide predictions about the future, scaffold explanations of what we observe, and drive the development of models to both explain and predict how the world works.

Predictions in science are not random guesses, but rather informed prognostications based on one's prior knowledge and evidence from one's prior experience. The ability to predict accurately what will happen in a new situation can be a strong indication of working logic. Explanations of the natural world are attempts to account for multiple lines of evidence from multiple scientists, and new discoveries and lines of inquiry are often born when apparently contradictory lines of evidence collide in the formation of an explanation. The National Research Council, in succinctly describing the Essential Features of Classroom Inquiry in *Inquiry and the National Science Education Standards* (NRC 2000), underscores the importance of evidence in developing explanations:

- The learner is engaged—mentally engaged—in a scientifically oriented question.
- The learner gives priority to *evidence* in responding to a question.
- The learner uses *evidence* to develop an *explanation*.
- The learner connects *explanation* to scientific knowledge.
- The learner communicates and justifies *explanation*.

One might argue that evidence is, in fact, the currency of knowledge in the scientific community. Reproducibility of evidence is considered by most scientists to be a core tenet of science. Unless scientists can support their ideas by providing evidence that can be confirmed by other scientists and lines of logic whose validity can be judged, their ideas are not likely to be accepted. Models—be they mental, physical, math-

SECTION II: Exploring the Triad Framework

ematical, or computer—are tools that scientists use to bring together large amounts of evidence into a singular, coherent representation that can both explain and predict phenomena in the natural world.

Yet, analyzing, interpreting, and using evidence is challenging. It's all too easy for even the most experienced scientist to jump to conclusions based on preconceptions, assumptions, and expectations. It is a skill and a significant challenge to firmly ground oneself in evidence when constructing multiple, possible explanations of how the world works. For many students the collection, examination, analysis, and evaluation of evidence is daunting. Often, students have opportunities to collect and record evidence during science activities, but their opportunities to consider the evidence and apply reasoning in the development of an explanation are constrained. Students tend to lack opportunities to analyze evidence and use that analysis to explain a phenomenon or construct a working model of the world. Model building—whether conceptual, visual, or physical—is rarely tied to evidence; the construction of a model is not based on evidence with which the student is grappling. Often, students are simply expected to connect their evidence with expert-generated explanations—from a book, or a teacher, for instance. In addition, the pace and timing of secondary classroom lab activities usually doesn't involve student reflection: time to examine and use evidence in order to make predictions about a related, future experiment or the critical discussion of evidence, what ideas it supports, and what it cannot explain (NRC 2007). Students are not skilled in the back-and-forth discussion and speculation of evidence that is so essential to the process of science, nor are students always comfortable with the risk taking necessary to speculate about what evidence means nor resilient in the face of the multiple attempts that are inevitably required to make sense of evidence.

Some strategies we found helpful in supporting students in using evidence include:

- Encourage students to make predictions based on evidence. If students can repeat related experiments and predict what will happen if they change a variable, they can begin to develop reasoning skills in building explanations. This can also be a valuable form of assessment.
- Have students synthesize multiple lines of evidence to support a claim or assertion. Environmental studies can provide good arenas to gather a variety of pieces of evidence and to compare alternate explanations.
- Work with students to understand the difference between observation and inference. For example, in a case where one observes a hissing cat, it would be an inference to assert that the cat is mean. But the cat may be a very nice cat who

is hissing because she thinks another cat will attack her kitttens.
- Use models in many different ways and acknowledge that they are not the real thing. We traditionally represent atoms as some sort of flat, miniature solar system because this model helps to explain our thinking about how atoms connect to one another. This is a good model for making predictions, but it is not a good model for explaining why atoms do what they do. Furthermore, although scientists have detected and prodded a single atom, they have never actually viewed one.

SECTION II: Exploring the Triad Framework

The Most Difficult

Analyzing data and drawing conclusions was probably the most difficult aspect of science investigations for our team to get to. The girls had incredible difficulty in supporting their ideas with data. Data collection was not always easy to do and not the "fun" part. Girls would want to leave before we could do a wrap-up of the activity, yet this was the time when we had planned to analyze the data and draw conclusions.

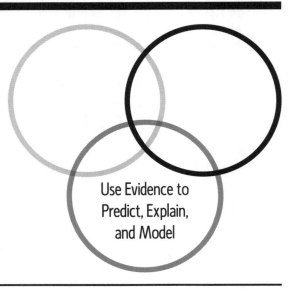

Use Evidence to Predict, Explain, and Model

Reflection Questions

- The author shares that the girls did not see data collection as the "fun" part. To what extent does this accord with your own experience? What aspects of a science investigation do your students most enjoy?
- It is not unusual for teachers to run out of time before the wrap-up discussion at the end of a science investigation, yet the discussion of evidence is a crucial part of doing science. What is gained during this discussion of a science activity following an investigation? What is lost if this part of the lesson doesn't happen?
- In the above question, it is assumed that discussions of evidence in a science classroom must occur at the end of an investigation. What drives this assumption? How could a lesson be structured so as to give students opportunities during their investigation to discuss emerging evidence with others?
- It is not unusual for students to assume the outcome of an experiment based on their knowledge from textbooks, experiences, or fellow students. However, some of the most important discoveries made in science have come from scientists taking a closer look at evidence from an experiment that has been done by dozens, if not hundreds, of other scientists before them. How do we focus students on the evidence from their own investigations, rather than rely on expectations based on what others have observed?

CHAPTER 4: Science Goals

Links

- Student Goals: We Have Reason to Believe
- Science Goals: To Trust in Their Own Logic
- Teaching Goals: I Could See How Much I Learned
- Teaching Goals: Resurrecting Socrates

This science team found analyzing data was difficult for them.

Girls in Science

SECTION II: Exploring the Triad Framework

Scientifically Dissatisfied

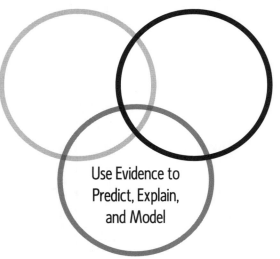

Our next activity was Vitamin C. This was one of the girls' favorites. I was scientifically dissatisfied because the results were all over the place. Two of the teams thought they had found Vitamin C in vinegar, which was obviously wrong to us. In the geology-focused making-mountains activity at a later club meeting, we had a similar problem. On stirring a clear cup of water containing sand, silt, and gravel, one should see a layer of gravel, then sand, then silt settle out. However, due to an experimental design flaw, the gravel was hidden by the sand, so the girls speculated where the gravel had gone.

Our problem in both cases was how to deal with the fact that the results were wrong. This was usually due to experimental design flaws or the girls' making a trivial mistake somewhere. We tried to encourage them to think critically and suggest explanations for their unexpected results, but what should we do if the evidence on which they were basing their conclusions was wrong? How can I as a teacher stand up and say, "Well I know vinegar doesn't have vitamin C, and therefore you did something wrong," when they have used experimentation and have some evidence that vinegar does have vitamin C? By the same token, I could not disprove the theory preferred by the girls that the gravel would sit in between a bottom layer of sand and a top layer of silt, because they described what they saw by looking through the side of the beaker. It was very frustrating to lead a discussion knowing they were wrong, but seeing that their conclusions were valid from the evidence put before them.

Reflection Questions

- Describe a time when you have done an experiment either yourself or with students in which the observed evidence did not match the expected outcome. Do you think this means the evidence was wrong? Explain.
- What might have been gained by explicitly telling these students what the ex-

pected outcome of the investigation was predicted to be? What might have been lost? When would you have told students that the expected outcome differed from their own results?
- How might you respond differently to students whose evidence is unexpected for experimental reasons than to students whose conclusions are unexpected due to a lack of attention to evidence?
- What are the potential effects on student learning of a teacher sharing that he or she is surprised by an experimental result and is not sure what caused the observed outcome?

Links

- Student Goals: Making Mistakes
- Science Goals: Answers Are Not the Goals
- Teaching Goals: Theory Is Easy, Practice Is Difficult
- Teaching Goals: Stop in My Tracks

SECTION II: Exploring the Triad Framework

You Can Lead a Horse to Water

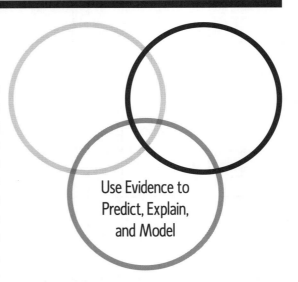

The crime-lab activity was new to me and something that I might try next year during my chemistry unit. It was quite exciting to the girls to have the crime introduced by an actual police officer. Our school is fortunate to have a police officer who frequents the building and was willing to introduce the crime and talk about the scientists who work with and for the police. With the stage set, we gave the students roles for this activity: recorder, materials handler, and facilitator. This helped somewhat to equalize the playing field for the students, but it did not prevent them from jumping to conclusions. We tried in various ways to lead them to logical reasoning, but they always wanted to bring in circumstantial evidence to defend their positions rather than sticking purely to the evidence from the crime scene. The suspects' motives quickly became the more interesting focus of the investigation. Two sessions of trying to motivate the girls to defend their positions with the evidence finally resulted in our leading them to organize their data in a way such that there was only one criminal, and they convinced each other of this. I think some of them were still more involved with the colorful inconclusive evidence. What I learned from this experience is that you can lead a horse to water, but you can't make it drink. Perhaps in our efforts to pique their interest we had stimulated their imaginations as well?

Reflection Questions

- Describe a time when you struggled to keep your students focused on the evidence at hand in a science lesson. What kinds of strategies did you use to guide them?
- In particular, when do you give students a structured way to record their evidence—such as a worksheet or data table? When do you allow each student team to decide how to record its evidence? On what do you base these decisions?
- How might a teacher talk explicitly with students about the differences between

CHAPTER 4: Science Goals

speculating about what may have happened versus scientifically engaging in the work of gathering and using evidence to draw conclusions?

Links

- Student Goals: We Have Reason to Believe
- Science Goals: To Help and Teach Each Other
- Teaching Goals: Theory Is Easy, Practice Is Difficult
- Student Goals: A Little Unnerving

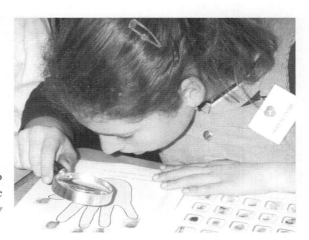

Getting students to focus on the evidence in a crime-lab activity wasn't easy.

SECTION II: Exploring the Triad Framework

Figure 4F

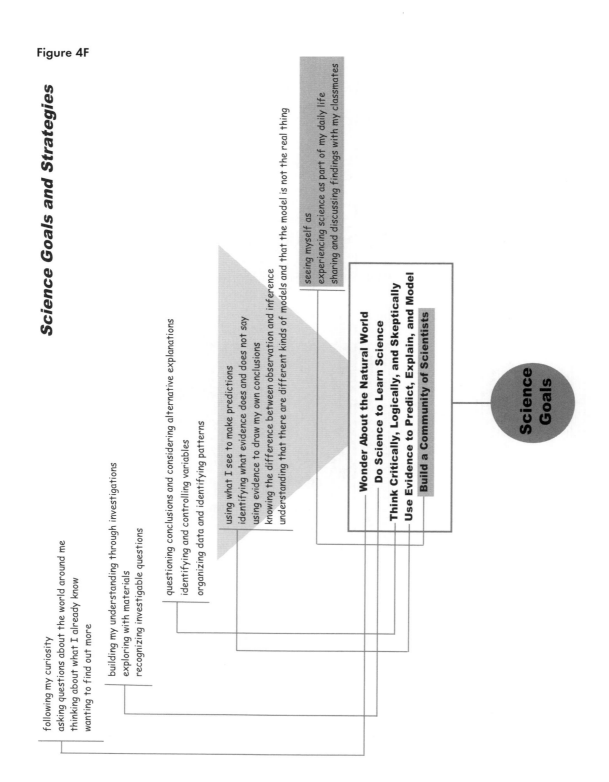

CHAPTER 4: Science Goals

Essay: Build a Community of Scientists

The eccentric white man with thick glasses and a white coat, alone with a smoking beaker in his laboratory, embodies the unfortunately popular view of how scientific discoveries are made. The development of scientific knowledge is, in modern reality, a collaborative endeavor carried out by a large community of scientists working in laboratories, universities, companies, and across subdisciplines and international borders. As a result, science, like many other ways of human sense making, is inherently social. We humans learn the most when we have others with whom we can examine assumptions, exchange ideas, share techniques, and experience challenges and triumphs. Scientists participate in several levels of scientific communities: within their own laboratories, within the cross-disciplinary community of their universities, and in the context of national and international scientific meetings. They engage in peer review of grants and publications, spirited discussions of how to interpret results, technical strategy sessions about the optimal approach to gathering data, and uncounted exchanges of wonderings over coffee and meals. If we aspire to have girls experience the joy and satisfaction of science as scientists do, then building communities of scientists in our classrooms is an essential task.

Yet, girls often experience and see science as a solitary endeavor laced with competition. They view success in science as available only to the smartest, not as accessible to everyone. And, finally, success in science can be viewed negatively among adolescent girls, potentially decreasing their social status among boys and attracting unwanted labels such as *nerd* (Sadker and Sadker 1994; AAUW 1998). Research on student perceptions of science have shown that science is viewed as a discipline fraught with isolation and competition, devoid of the personal, the creative, and the relevant. In fact, some researchers have attempted to define what the nature of science is and how it can be most authentically portrayed in schools (Lederman 1992). What has emerged is a framework that emphasizes science as a tentative and dynamic process, as the product of human inference, as requiring creativity and imagination, and as heavily embedded in the culture and society we live in (Lederman 1992). Seeing science as tentative rather than absolute, as valuing creativity and not solely objectivity, and as driven by questions that come from our personal, family, and cultural experiences has the potential to open the doors of science to girls and a more diverse population of students in general.

Having students see themselves as scientists and experience being members of a community of scientists are fine ideas, but how do we make it happen? What would it look like in a classroom? Many girls have not had the chance to experience the joy of collaboration with others on a problem or a community of students who are committed to finding answers, supporting and questioning one another, and collaboratively putting it all together. Girls may not have seen themselves as successful in science, much less leaders in science, or perceive themselves as science people. The science experienced by students may be constrained by the educational system, standards, and curriculum, and as a result, may leave many students, not just girls, questioning the relevance of science to their own lives. Given more opportunities to ask their own questions, do experiments that they genuinely care about, and lose themselves so much in their own questions and the drive toward understanding that science is no longer an academic exercise, but a quest for insight, girls may begin to see science as relevant to themselves, see themselves as capable, and, most important, have the desire to ask more questions. Perhaps through seeing herself as a scientist and a critical member of a community of scientists, a girl may begin to integrate science into her personal identity.

So, how can we as educators strive toward creating a genuine community of scientists in our classrooms where students can begin to see themselves as scientists? In many ways, the first steps toward building a community of scientists are suggested in the other Triad Science Goals. To build a community of scientists, we as teachers can provide opportunities for our students to do science to learn science in open-ended, student-driven investigations. We can push and guide our students to think critically, logically, and skeptically about their investigation results, their classmates' results, TV shows, and the news, and foster an ongoing conversation among students that is at once respectful and supportive while it is skeptical, questioning, and in search of the collective best explanations. We can aspire to create a classroom culture in which it is strikingly ordinary and common to share what fills us with wonder about the natural world, to question one another in a supportive way, and to dream up models that can predict and explain what we observe. Building such a community of scientists does not happen either quickly or by accident, but rather emerges with its own set of values, ways of interacting, customs, celebrations, and style. It means believing that you and your fellow community members—teachers and students alike—can ask questions, solve problems, disagree, have insights, and together create new knowledge. A willingness to develop collaboratively this community of scientists with students is indeed a brave act on the part of any teacher.

CHAPTER 4: Science Goals

The Strength of the Group

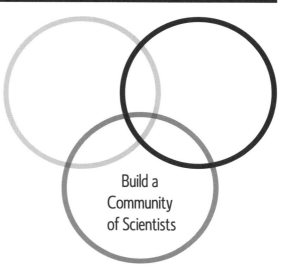

This strategy of splitting the entire club into smaller groups was also effective for allowing girls who didn't typically socialize with one another to work together. Often times, this paired girls from different ethnic groups. I believe that this increased the strength of the group, because everyone began to better know their science club as a whole and to feel camaraderie. This was particularly evident when our group did a lamb-heart dissection. The girls were asked to orient the heart and figure out the flow of blood within the chambers of the heart. This was not necessarily an easy task, and some girls grew frustrated in the process. I found that the girls who had successfully traced the flow of blood were then equally happy to begin giving other girls at their table hints to help them succeed as well.

Reflection Questions

- This scientist makes a strong connection between assigning students to groups and the growth of community within this club. Do you agree that this example illustrates such community? Why or why not?
- Do you believe it's important for students in a classroom to get to know and be required to work with as many other students in the class as possible? Explain.
- Describe a time in your own learning in which you were assigned to a small group. What was positive about this experience? What did you find challenging about it?
- It is not uncommon for students to express a preference for choosing their own seats in a classroom and working alone. What is it about the typical small-group experience in the classroom that might cause some students to question them as a valuable learning structure?
- Given the above, what kinds of things do you think it is important to consider and plan for in assigning groups in your classroom?

SECTION II: Exploring the Triad Framework

Links

- Student Goals: Where to Draw the Line
- Student Goals: She Wanted to Do It Herself
- Teaching Goals: To Cunningly Mediate Equity
- Teaching Goals: See What Happens
- Teaching Goals: Back in the Classroom

Student Work

5. Name one person you worked with whom you didn't know very well before the club. Describe how you worked together. I worked with Marie she was a good parner and she an I helped each other with the gears. She helped me with giving me ideas. And it was fun.

CHAPTER 4: Science Goals

When Science First Was Really New

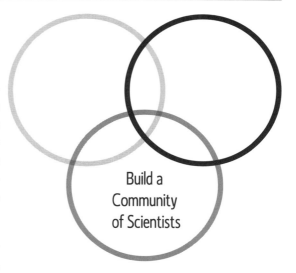

This vignette is taken from a group interview with eighth-grade students in an all-girls club. They have talked about why they joined the club and what it is they think that scientists do. This next probe asks them to compare similarities of what scientists do with what happens in clubs.

Jenna: Um, what we do doesn't really seem like what scientists do. Cause when I picture scientists, I think of them working in laboratories, um, and doing research, and collecting data.

Claire: I think it's like when we did the little crystal things and tried to make crystals by evaporating the water.

Isa: Yeah I agree, because, we got to look at them, and see how it developed after a while, with magnifying glasses, and, I …

Ilene: [Interrupting] I think that scientists probably don't do that now, but when they first started, when science first was really new, when people were like interested in it. Like the Vitamin C project, they probably don't do that anymore, cause they know, but, before, they were probably testing for things and that's how it was found out.

Lucy: Yeah, I also think the squeeze-box activity, with the different layers of substances, was something that scientists would probably have done before to figure out how earthquakes impact different areas.

Angel: I think that everyone has to start somewhere, and Triad may not be like what scientists are doing now, trying to discover things, but it's how they started.

Diana: I think also things that the science club has been doing are something that scientists would do maybe for fun or their own interest … but I think they do mainly things that would help people, cure diseases or something. Like the crystal project, if that was going to help somebody.

Girls in Science

SECTION II: Exploring the Triad Framework

Reflection Questions

- To what extent do these students see themselves as scientists? What do you think influences their views?
- Diana says, "things that the science club has been doing are something that scientists would do maybe for fun or their own interest." What attitudes about scientists and science are evident from this comment? How are Diana's ideas similar to or different from your own?
- The girls refer to their own science learning experiences as similar to when science "first started" and "when science first was really new." What do you think they mean by these statements?
- Is science always about constructing new knowledge for the discipline, or is it also about constructing new knowledge for oneself?
- Scientists doing research do not know the answers to their experimental questions. Do you think not knowing the answer is important for students doing investigations in classrooms? Why or why not? To what extent does knowing that you can find an answer in a book or on the internet discourage you from trying to find answers for yourself?

Links

- Science Goals: Nobody Knows What's Inside
- Science Goals: Not What We Had Planned
- Teaching Goals: I Could See How Much I Learned
- Teaching Goals: The More We Expected

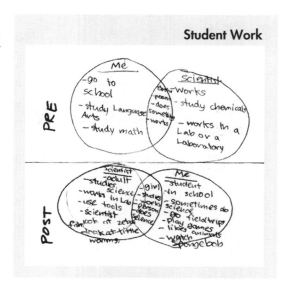

Student Work

CHAPTER 4: Science Goals

Have to Make It Right

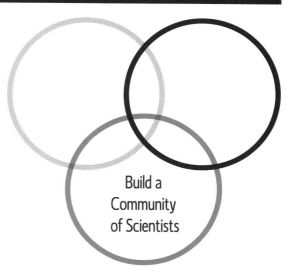

This vignette is from a group interview with sixth- and seventh-grade students in an all-girls club. They have talked about why they joined the club, and what it is they think that scientists do. Next they compare the similarities and differences of what scientists do with what happens in clubs.

Interviewer: So when you guys have been talking about what scientists do, you've mostly been talking about things like making the world a better place, and getting rid of viruses, and also about things exploding, and chemicals—is what you do in science club—how is that connected to, or like, what you think scientists do?

Sylvia: Well, we do experiments to find out what things are like, and it's just a fun way of finding out things. It's cool!

Melanie: When we're doing experiments, we, like, learn what happens in the world, like about earthquakes and volcanoes, and um, we have fun, especially because we make a mess, except at the end when we clean it all up.

Interviewer: So how are the experiments you do different than what scientists do?

Melanie: Well, because, the experiments that we do, they're like more safe, we could, like, if it doesn't work out we could probably try again. But, with scientists, the experiments are more dangerous. They have like protective gear so they don't like get poisoned or something. So in science club we don't have to wear much protective gear and um we…

Sylvia: We don't have to be too careful of stuff, and, in science club, it's basically just about having fun, and, if you're a scientist, you just have to make it right.

Interviewer: How can they be sure to make it right?

Sylvia: Well, everyone makes mistakes, but, well, they study more, I guess, and [pause] but we don't have to make everything right, and it's part of their job to, I guess, I don't know, do the right thing.

Girls in Science 163

SECTION II: Exploring the Triad Framework

Melanie: What she said, I do agree, but I think, like, first they, like, write on a piece of paper what they do, and then they just, like, follow step-by-step to, um, try to make sure that they don't mess up and things, and so then what they're trying to accomplish is to make it on the first try. So they don't have to do it over and over and over.

Tara: It's like what scientists have to do, like, um, they want to make the world better, and they like can only study much, and yeah. And like we don't have to make seriously…

Melanie: That's what is cool about science club, that even though we have things we should be serious on, it's still fun in a way.

Reflection Questions

- What do you think these students mean by saying that scientists "have to make it right" and "do the right thing"?
- The students above also perceive science as dangerous and serious. What do you think are the origins of these student ideas about science?
- How do these students' perceptions of science keep them from seeing themselves as scientists? Do you view their perceptions as correct or incorrect? Why?
- What might be done in the classroom to help all students believe that they are scientists whenever they are doing science?

Links

- Student Goals: Not Having Step-by-Step Instructions

- Student Goals: We Have Reason to Believe

- Science Goals: Not What We Had Planned

- Teaching Goals: Talking in Questions

> A student said to me "Now I take science class more seriously because I didn't know that science could be so fun." It's so striking to me that her Triad experience influenced her attitude toward science class, not to mention that most people wouldn't predict that something "fun" would make a kid more "serious."
>
> —*Triad Teacher*

CHAPTER 4: Science Goals

To Help and Teach Each Other

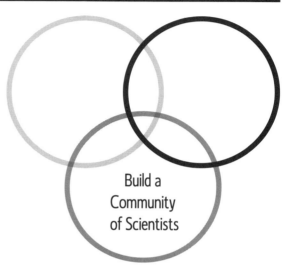

I learned how important body language can be when you are teaching. By walking around to each table, we could show that we were there to help the girls if they needed us, but, by keeping our hands in our pockets, we were showing that we expected them to do as much on their own as possible. One good example of this is the crime lab activity, for which we used two of our club meetings.

In the first meeting, we taught the girls four forensics skills—fingerprinting, mystery-powder analysis, ink chromatography, and hair microscopy. In the second meeting, the teachers and scientists became suspects of a crime, and the girls were the detectives with the whodunit on their hands. Each of us proclaimed ourselves innocent and the other three Triad club sponsors guilty. As a result of our role-playing, the girls couldn't trust us to help them with the four skills. Instead, they needed to help and teach each other the activities to solve the crime.

Reflection Questions

- How might their inability to rely on their teachers and scientists have affected the way these students approached this science activity?
- What are other practical strategies you use to build a community of scientists in your classroom?
- At the end of this activity, how might telling or not telling these students who the real criminal was have affected their sense of community and self-reliance?
- What kinds of lessons in your classroom encourage students to rely on one another for help? What other science lessons in your curriculum could be restructured to do so?

SECTION II: Exploring the Triad Framework

Links

- Student Goals: We Have Reason to Believe
- Science Goals: You Can Lead a Horse to Water
- Teaching Goals: A Different Role
- Teaching Goals: No One Felt Uninvolved

CHAPTER 4: Science Goals

Turning to Nadya

This vignette is from notes written by an observer at an all-girls after-school science club for sixth, seventh, and eighth graders. Due to a power loss at the time of the meeting, the Triad teacher and scientist team were forced to come up with a new activity on very short notice. The original plan was for the girls to use the internet to learn more about comets and the solar system. When the power went out, the activity was altered so that small groups of girls were asked to answer a question based on their prior knowledge.

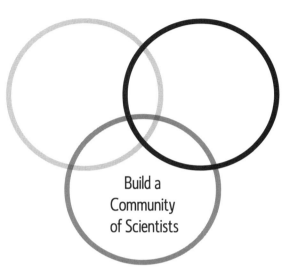

I've watched this club wrestle this year with the difference in science knowledge among girls in the club. Often, other girls would turn to one girl—Nadya—to answer questions no one else could. The adults recognized her as particularly able, too. On this day, as the scientist was assigning questions about comets to groups of girls, he added a challenge question, "When is the comet's tail longest?" to the question "Why does a comet have a tail?" This proved to be a good activity for Nadya and Fanny, because they had enough background knowledge to really grapple with both parts and because it was relatively easy for the adults to help them work through their partially formed, very scientific, ideas. But it was harder for the adults to help girls who had much less knowledge, because these girls had fewer ideas to discuss, even when the questions were much easier. In that way, despite the adults' attempt to level the playing field by giving Nadya and Fanny a harder question, this activity really showed the differences in ability between girls. And the girls couldn't help but notice it during the time they shared their posters.

Reflection Questions

- How was sharing and discussing ideas in this lesson potentially a barrier to having students feel like a community of scientists?
- How does the nature of the questions students are given contribute to emphasizing differences in background or ability among the girls? How might you change the questions used in this lesson?

Girls in Science

SECTION II: Exploring the Triad Framework

- Describe a time in your own experience as a learner in which it was obvious that students in the group had very different backgrounds and some had much more to contribute than others. What was this experience like for you?
- Describe a classroom activity you have seen or done in which students with varying backgrounds all were able to contribute significantly to the group's learning. What do you think it was about this activity that made this possible?
- What kinds of factors tend to contribute to our assessments of students' abilities? To what extent are these factors truly about potential to learn versus other attributes that may be more directly associated with life experience, class, race, sex, or culture?

Links

- Student Goals: I Shouldn't Have Come
- Student Goals: I Assumed That Our Girls Would Feel Comfortable
- Science Goals: Answers Are Not the Goals
- Teaching Goals: No One Felt Uninvolved
- Teaching Goals: Science Is Not a Priority for These Students
- Teaching Goals: Many People Got a Chance

> Many girls were reluctant to share their ideas with their clubmates. I attribute this reluctance partially to our school site. There was a pronounced desire to be correct (or not to be wrong) among our girls. We tried to stress that we didn't care about the factual accuracy of their ideas but only that they try to express them.
>
> —*Triad Teacher*

CHAPTER 5:
Teaching Goals—Striving for Gender-Equitable Science Teaching

SECTION II: Exploring the Triad Framework

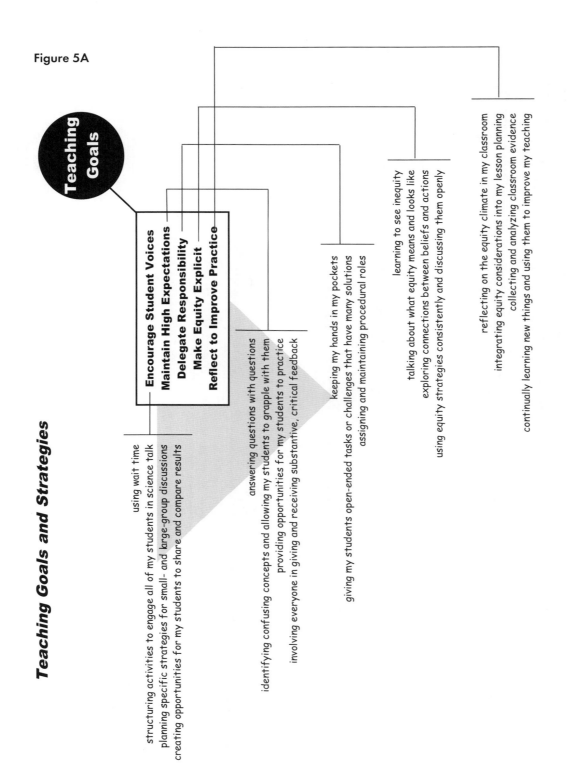

Figure 5A

Teaching Goals and Strategies

Teaching Goals Table of Contents

Introduction to the Chapter 173

Essay: Encourage Student Voices 177

    I Watched in Awe 180
    The Quieter Girls 182
    I Have to Introduce Triad 184
    See What Happens 187
    Many People Got a Chance 188

Essay: Maintain High Expectations 191

    The More We Expected 194
    Talking in Questions 196
    Can You Help Me? 198
    Answering Student Questions With Questions 200
    Theory Is Easy, Practice Is Difficult 202
    Science Is Not a Priority for These Students 204
    Accepting Stereotypes 206

Essay: Delegate Responsibility 209

    Keeping Your Hands in Your Pockets 212
    Does This Bridge Look Better Than It Did Last Time? 214
    I Could See How Much I Learned 216
    A Different Role 218
    No One Felt Uninvolved 220

Essay: Make Equity Explicit 223

    To Cunningly Mediate Equity 227
    Like Dad 229
    Talking About Equity 231
    Way Beyond Our Expectations 233
    Stop in My Tracks 235
    Anyone But the Boy 237
    My Own Tendency 239

SECTION II: Exploring the Triad Framework

Essay: Reflect to Improve Practice 243
    Back in the Classroom 246
    By Scoring When a Girl Participated 248
    Personal Development...250
    Resurrecting Socrates...252
    At First I Was Hesitant 254

Introduction

Triad Teaching Goals
Encourage Student Voices
• using wait time
• structuring activities to engage all of my students in science talk
• planning specific strategies for small- and large-group discussions
• creating opportunities for my students to share and compare results
Maintain High Expectations
• answering questions with questions
• identifying confusing concepts and allowing my students to grapple with them
• providing opportunities for my students to practice
• involving everyone in giving and receiving substantive, critical feedback
Delegate Responsibility
• keeping my hands in my pockets
• giving my students open-ended tasks or challenges that have many solutions
• assigning and maintaining procedural roles
Make Equity Explicit
• learning to see inequity
• talking about what equity means and looks like
• exploring connections between beliefs and actions
• using equity strategies consistently and discussing them openly
Reflect to Improve Practice
• reflecting on the equity climate in my classroom
• integrating equity considerations into my lesson planning
• collecting and analyzing classroom evidence
• continually learning new things and using them to improve my teaching

The Triad Teaching Goals and their supporting strategies were articulated to aid teachers and scientists in seeing the role of their own behavior in creating equitable science learning environments for girls. For our community of teachers and scientists, it was a challenge to direct our gaze away from girl students and toward ourselves as causal sources of the gender inequities we had come to recognize in our classrooms. Many of us had not examined our differential interactions with students (Sadker and Sadker 1994; AAUW 1998), nor had we considered how our own choices in structuring a science lesson could produce equity or inequity depending on its design. And we had certainly not entertained the possibility of explicitly talking with girls and boys about issues of gender equity and the strategies we were using to try and address it. It became strikingly clear that, if we wanted girls to develop the skills and behaviors outlined in the Triad Student Goals, then we also had to learn new skills and behaviors that would help us become more gender-equitable science teachers. But what were those skills and what could we practically do differently in our classrooms to make them more equitable? In drafting the Triad Teaching Goals, we have certainly stood on the shoulders of giants. The Triad Teaching Goals were informed by the teaching standards articulated in the *National Science Education Standards* (NRC 1996), by research that indicates that talking on task while manipulating educational materials leads to greater learning gains (Cohen and Lotan 1997), that strategies such as wait time increase student participation (Rowe 1987), that teachers' expectations of students affect learning gains (Rosenthal and Jacobson 1992), and that interactions with students can build or hamper self-confidence (Gordon 1995). As a community of teachers and scientists, we asked ourselves what this research implied *we should be doing* as adults to construct an equitable-classroom learning environment and what *we should actually not be doing*. What emerged was an ideal that we all strived toward, a picture of an educator who is skilled in practicing wait time, who designs all lessons with an eye toward engaging every student, and

SECTION II: Exploring the Triad Framework

who is explicit with her students about the equitable teaching strategies he or she uses and why. Our ideal educator self is confident enough to keep his or her hands in his or her pockets, to guide both girls and boys verbally through challenging, hands-on experiences, and to delegate responsibility to students during science lessons through assigning group roles and maintaining them. In the ideal, we practice restraint and do not answer student questions immediately, but rather respond with questions to guide them in plumbing the depths of their own knowledge before attempting to add new information. In fact, the Triad Teaching Goals are often reminding us that *not doing things for students*—waiting, not always answering questions, not always fixing things that aren't working—is a critical idea in building strong students in science, especially girls. Throughout this chapter, you will find more information on each of the Triad Teaching Goals—*Encourage Student Voices, Maintain High Expectations, Delegate Responsibility, Make Equity Explicit, and Reflect to Improve Practice*—highlighting the importance of these teaching skills and behaviors in science and how our own community has struggled to reach them.

SECTION II: Exploring the Triad Framework

Figure 5B

Teaching Goals and Strategies

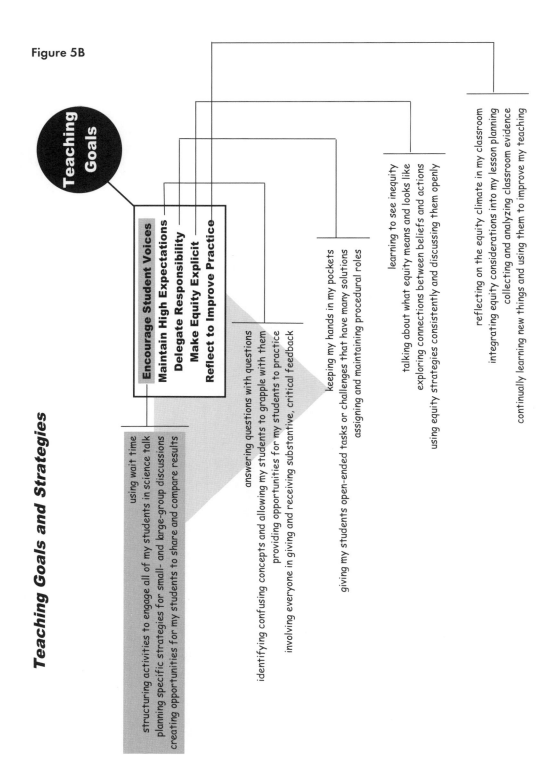

CHAPTER 5: Teaching Goals

Essay: Encourage Student Voices

In science gaining new knowledge comes not only from making observations and doing investigations but also from the talk that surrounds these activities. For scientists, sharing their findings and articulating their interpretations is as important as conducting the investigations. In addition, they need to be able to listen carefully and question their colleagues' results in order to further everyone's understanding. In light of this, student success in science—especially for girls—requires that students have opportunities to find and develop their voices. Encouraging students to use their voices and express their views, questions, and confusions is essential to developing a classroom in which students are actively engaged in science learning.

Research on classroom interaction patterns has shown that, in general, these patterns favor boys' voices over girls'. Boys express more self-confidence and assertiveness than do girls (AAUW 1998). In a heterogeneous environment, boys dominate discussions and are more likely to interrupt girls. If girls do interrupt, they are more likely than boys to be reprimanded (Holden 1993). In coeducational groups girls are also likely to talk at a less-abstract level than boys (Martinez 1992). These patterns are encouraged and maintained by teachers' behaviors: Teachers ask higher-order questions of boys, are more likely to call on boys first, give boys more praise and substantive feedback, and, when the task is the same, will provide instructions for boys but show girls how to do it (Tobin and Garnett 1987). As a result, boys are more likely than girls to have higher numbers of concrete experiences, more experiences in problem solving, greater opportunities to process their experiences through verbal communication, and more acknowledgment and feedback about what they have done. Involvement of this nature fosters the boys' continued interest and study in science. Most critically, these teacher expectations and interactions promote self-confidence and a richer abstract understanding of theoretical concepts related to concrete experiences. Girls, who tend to be excluded from this spiral of positive reinforcement, may become discouraged. They receive less positive stimulation and suffer from loss of confidence because of their lower level of participation and acknowledgment. As a result, they participate even less. Relevant to these gender-based discrepancies is a large body of work that has shown that verbal interaction with peers during the kinds of open-ended group tasks that are authentic science result in significant learning gains (Cohen and Lotan 1997). When classroom interaction patterns result in girls talking less, they also result in girls learning less.

Students come to our classrooms with their own inclinations about talking in front of their teachers and peers. There are those who are very hesitant to speak regardless

SECTION II: Exploring the Triad Framework

of classroom climate. Others are happy to engage in one-on-one public dialogue with teachers and peers alike. A large segment in the middle would be happily willing to speak if their classrooms were structured in a way that invited them to share their thinking as part of a collaborative classroom quest for understanding. The question for us as teachers is how to go about engaging all of these students in productive science talk.

In Triad we all struggled to change our teaching behaviors and classroom structures that contributed to the following patterns: our teachers' voices being preeminent, a majority of the conceptual interactions occurring between student and teacher, a few student voices dominating, and students having limited time to clarify and share their thoughts with one another. These changes were not easy, but we did identify some helpful strategies that might be useful to you as well:

- Make a distinction between the times you need to present and explain a set of concepts and the times you encourage and expect students to express their own understanding and ideas. For students this includes meaningful opportunities to think about and express their ideas both as they formulate them and as relatively finished conceptual thinking. It necessitates not only that students speak but also that they listen carefully and express thoughts on paper in text, drawings, and graphs.
- Be attentive to the participation rates of individual students. Finding ways to collect data on which students talk and under what conditions can give you insight into your individual students and your own teaching. It can be difficult to do this while facilitating a discussion, so you may wish to enlist some help from a parent or colleague.
- Employ teaching strategies that allow for student thinking time. Increase your wait time—the seconds of silence given students between a teacher-posed question and a call for a response. A wait time of three seconds greatly increases student participation, quality of responses, and divergent thinking (Rowe 1987). A silence of three or more seconds also has been shown to be supportive of student thinking as they share their thoughts, after they share a response, after the teacher presents information, and as they are completing a task (Stahl 1194). Students can also write or draw their ideas before sharing them verbally, or they can talk with a partner or in small table groups before sharing with the whole class.
- Encourage multiple voices and new voices to contribute to class discussions. Use a talking stick and talking circle to encourage all to participate, and give students opportunities to finish their thoughts uninterrupted. Use a limited

number of talking chips per student to equalize participation. In the whole class or large-group setting picking popsicle sticks with students' names on them can help you ensure that you are giving all of your students the opportunity to express their ideas. As teachers, we can make it clear that everyone has important thoughts to share and that we expect to hear from everyone. During open discussions we can return to those students who have not spoken and ask them to express their thoughts.
- Be intentional about how you structure your lessons. When students are engaged in open-ended investigations and are working together as a community of scientists to build their knowledge of a particular set of phenomena, every pair or group has observations and thoughts to share. As students talk with one another about their questions, approaches to investigation, and results, they develop both their scientific and general communication skills in addition to learning about the topic at hand.

These are just some places to start. What is most important is that we acknowledge that student talk plays a vital role in their science learning and we each commit to doing all we can in our specific contexts to encourage all of our students' voices.

SECTION II: Exploring the Triad Framework

I Watched in Awe

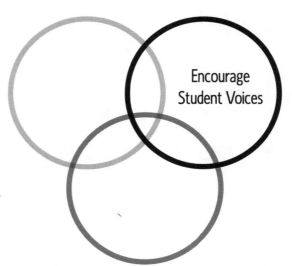

I can remember the Triad professional development meeting where we focused on the issues of gender equity for the first time. I had been relatively unaware of research on the tendencies of boys to overshadow girls in science classrooms. The next day, I watched almost in awe as my classroom modeled the behavior we had discussed. I began to take an active role in balancing the participation of my students. I would alternate between boys and girls to answer questions. I found that often times girls would raise their hands but would go unnoticed because the boy in the front of the room would be waving his hand around and jumping out of his seat. I also found it important to call on girls even if they were not volunteering. It became clear that, once they had answered a question correctly, they were more willing to volunteer an answer on the next turn.

Reflection Questions

- What are some of the observations that this teacher made about equity of student verbal participation in his or her classroom? What observations have you made about how often boys and girls speak in your own classroom?
- How might you collect evidence and analyze the relative participation of girls and boys in your classroom? Of students from different cultures? With different language abilities?
- What strategies do the teacher in this account use to encourage participation by girls in her classroom? How would the strategies employed by this teacher work with your students?
- What are some other strategies that you have successfully employed to encourage participation by quieter students in your classroom?

CHAPTER 5: Teaching Goals

Links

- Student Goals: She Wanted to Do It Herself
- Science Goals: Above a Whisper
- Teaching Goals: My Own Tendency
- Teaching Goals: Stop in My Tracks
- Teaching Goals: By Scoring When a Girl Participated
- Teaching Goals: Back in the Classroom

> At the beginning of the club today, one boy asked a question, five times boys called out answers, three times boys answered questions, and one time a girl answered a question.
>
> —*Triad Teacher*

SECTION II: Exploring the Triad Framework

The Quieter Girls

This vignette is from a reflection written by a scientist who was a leader for an all-girls Triad club.

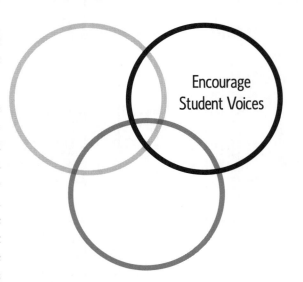

It was also a great learning experience to determine what types of group discussions encourage student voices and a high level of participation. We found that small-group discussions tend to promote the highest level of overall participation and seem to encourage the girls who tend to be more quiet to speak up and voice their ideas. We implemented these group discussions in a couple of ways—by posing questions to the girls and then asking for answers by asking them to raise their hands or going around the table and asking for answers from each girl in turn. By scoring when a girl participated in the conversation, we found that the level of participation was very good in either format. The quieter girls spoke up less in the voluntary setting, as opposed to the round-robin discussion. This helped me to understand the idea of equitable teaching—how it doesn't fall only into a category of gender or race and how to better incorporate it into each club meeting.

Reflection Questions

- In what ways was this team experimenting with different teaching methods to encourage students' voices?
- What have you noticed about your own students' verbal participation? What patterns have you noticed among your students?
- How do you most often engage your students verbally in your own science class: in pair discussions, table discussions, or whole-group discussions? Why?
- Research shows that wait time—waiting for several seconds after asking a question before calling on a student or asking another question—increases both the number of students participating and the quality of their responses to questions. Have you investigated your own wait time or attempted to lengthen it?
- To what extent do you believe that it's important for all students to talk as part of their learning? Please explain.

CHAPTER 5: Teaching Goals

Links

- Student Goals: I Assumed That Our Girls Would Feel Comfortable
- Student Goals: A Little Unnerving
- Science Goals: The Strength of the Group
- Teaching Goals: To Cunningly Mediate Equity

Student Work

4. Before being in the Triad Science Club, I thought science was <u>hard</u> <u>and scare</u> because... like sometimes you have to say out loud.

Girls in Science

SECTION II: Exploring the Triad Framework

I Have to Introduce Triad

This transcript was written by an observer visiting a Triad club. The setting is an all-girls, after-school science club. A team of five seventh graders is preparing to present at a conference that promotes math and science for middle school girls. In order to get some practice, the girls had acted as facilitators for the club activity for the day. This conversation is part of the wrap-up. Lucia, a veteran Triad student, conducted an informal survey to gather information that will help her explain Triad when she introduces Triad at the conference.

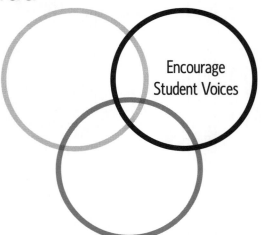

Lucia moved to the overhead at the front of room and wrote "Triad" on the transparency as a header and then directly below it wrote the question "What does Triad mean to you?"

Lucia: My question is …

Teacher 1: Why are you asking them this question?

Lucia: Because I have to present at the Expanding Your Horizons conference, and I have to introduce Triad and tell everyone what it means. I have a general definition, but I want to know what you think.

Lucia then proceeds to write responses on the transparency, while naming each respondent.

One student says, "Fun experiments." Someone whispers, "Science." Another squeals, "Just for girls!" Someone else says, "Girl power."

Lucia: OK. I need more ….

Mary: OK. Let's go around and each person say one thing about Triad.

Ann: Do stuff that boys can do.

Lucia keeps writing on the overhead. She waits quietly for another answer. The whole group is still silent and attentive. She asks the next student, "Need to think more?" and then moves on to the next student. She addresses each by name, saying those she knows, asking others what their names are. She keeps writing down answers.

Jennifer: Fun things after school.

Kate: Learning new things.

Lucia then asks the teachers and scientist for their thoughts:

Teacher 2: I think it means teamwork.

Teacher 1: Opportunity to work with real scientists and meet real scientists.

Scientist: Well, Triad, I'm thinking is three kinds of something. Three different kinds of people are here, know what they are?

Mary: Students, scientists, and teachers.

Lucia continues to write:

Jade: Doing different stuff with scientists, science with girls (we rule!).

Yung: Using different stuff while having fun.

Lucia then reads the whole list. All the students are engaged and listening.

Reflection Questions

- What strategies did these girls use to encourage their fellow students to speak?
- How do you think these students were prepared for these roles as facilitators of a discussion among their peers? How would you prepare your students for such roles?
- What strategies do you use to encourage verbal participation from all of your students?

Links

- Student Goals: I Assumed That Our Girls Would Feel Comfortable

- Science Goals: To Help and Teach Each Other

- Science Goals: The Strength of the Group

- Teaching Goals: See What Happens

CHAPTER 5: Teaching Goals

See What Happens

This vignette is from an observer's field notes of middle school girls at a Family Science Night.

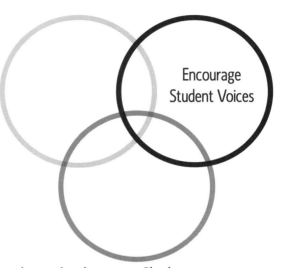

At this Triad Family Night, students led their families in science activities. Two girls, both sixth graders, were in the hallway playing with hoopsters, a type of model plane constructed from drinking straws and hoops made of card stock. One girl gave me some recommendations, such as one hoop attached to the straw should be bigger than the other. When I asked about variations, she suggested I try them out and see what happens. She kept encouraging me, while noting how far her own hoopster flew. Finally, I got mine going okay. As I left the hall, the two girls were setting up a tape line for people to line up at and fly their hoopsters as far as they could.

Reflection Questions

- What teaching strategies is this girl using?
- In what ways have you seen your own teaching strategies used by your students?
- What can students learn from teaching others the science they have learned? How can this encourage them to use their own voices in science?
- As a teacher, how do you feel when students ask you the same kinds of questions you ask them?

Links

 Student Goals: Not Having Step-by-Step Instructions

 Science Goals: The Strength of the Group

 Teaching Goals: I Could See How Much I Learned

Many People Got a Chance

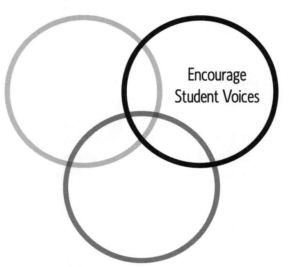

Much of what I learned about equity this year in Triad, I learned from observing the Triad staff. I appreciated how, at the retreats and professional-development workshops, they modeled the behaviors that they wanted us to use in the classroom. They made sure that many people got a chance to voice their opinions by keeping track of who had been called on and by making explicit the order that they would call on people when hands were raised. I believe their strategies encouraged people who might otherwise have remained silent to contribute to the discussion. I had a deep appreciation for this approach, comparing it to experiences that I had with group discussions in undergraduate classes in college or in lab-meeting or seminar-type settings in graduate school and beyond. I have always had a tendency to notice who is talking and who is not and to try to get the more quiet people to speak. I will definitely aspire to apply these techniques in my own classroom when I (hopefully) become a professor at a liberal arts college. Although I appreciated seeing the Triad staff apply equity guidelines with us, I found it much more difficult to apply them myself in the context of my Triad team and club.

Reflection Questions

- What strategies are described for supporting an equitable learning environment? What are other possible strategies?
- What strategies have you successfully used to encourage participation from all of your students?
- When have you had difficulty applying equity guidelines? Explain.

CHAPTER 5: Teaching Goals

Links

 Student Goals: We Have Reason to Believe

 Science Goals: Above a Whisper

 Teaching Goals: My Own Tendency

Triad staff modeled behaviors they wanted teachers to use in the classroom.

SECTION II: Exploring the Triad Framework

Figure 5C

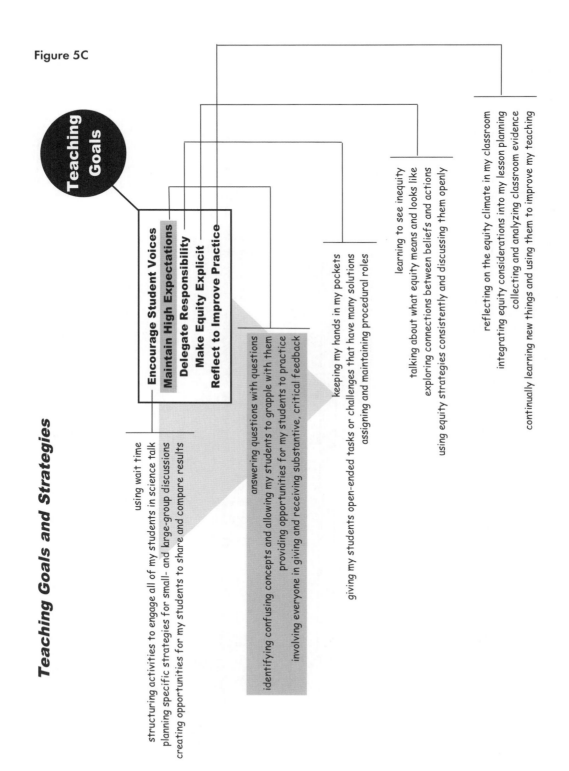

Teaching Goals and Strategies

CHAPTER 5: Teaching Goals

Essay: Maintain High Expectations

In our society, science is viewed by many as a field open to only the best and the brightest. We are led to believe individuals are born with particular aptitudes, gifts, and talents. These beliefs color our own and others' ideas about who is and can be a scientist, ultimately limiting the number of individuals entering the field and the diversity of perspectives that are represented. When classroom experiences leave us feeling as though we are not good at science, we lose confidence in our ability to learn and find it difficult to persist through the confusion, struggle, failure, and hard work that are at the core of solving a scientific problem. By extension, we tend to see scientists as the gifted ones, the experts, the ones who know. The culture both relies on and mistrusts the scientist, who is set apart from and above the average citizen. (See Science Goals: Walking Encyclopedia, p. 140) This is all about expectations—expectations about what it means to be smart, about what it takes to do science, and about students' abilities to learn.

Several different bodies of research demonstrate the important roles expectations play in student learning and performance. While the literature related to this topic is huge, we will focus on three particular aspects: theories of intelligence, stereotype threat, and tracking. A growing number of theorists and researchers are questioning our societal views of intelligence and their impact on learning (Dweck 2000; Resnick 1995). They suggest a shift from the belief that intelligence is innate to a view of intelligence as a set of characteristics and skills that are developed through effort on the part of the learner. Carol Dweck and her colleagues have been investigating the effects of praising students for their effort rather than their smarts and explicitly teaching students that intelligence is malleable. The evidence is convincing that this kind of input enhances both student motivation and performance (Mueller and Dweck 1998). As Dweck says, "Emphasizing effort gives a child a variable that they can control. They come to see themselves as in control of their success. Emphasizing natural intelligence takes it out of the child's control, and it provides no good recipe for responding to a failure" (Bronson 2007).

In another line of research, stereotypes and the expectations associated with them have been shown to have powerful negative impacts on performance. When students are told a group with which they identify—for instance females or African Americans—generally does worse on particular kinds of tasks or tests, they themselves do not perform as well as they do when the task is disassociated from the stereotype (Steele 1997; Steele and Aronson 1995). In the research literature, this phenomenon

SECTION II: Exploring the Triad Framework

is referred to as *stereotype threat*. The effects of stereotype threat can be reduced when students are told explicitly that the skills in question are not dependent on an innate fixed ability or when they are taught to think more generally of intelligence as something that can be developed through effort (Aronson et al. 2002).

Finally, research on tracking has shown that low-income students and students of color are disproportionately tracked into lower-level classes (Oakes 2005). In the case of a black female, these judgments may be based more on a student's adherence to white middle-class teachers' gendered expectations of what makes a good student, such as expecting that excellent female students are quiet, compliant, and timely with their work, than they are on the individual's interests and educational needs (Brickhouse et al. 2000).

As teachers, we are aware of the importance of having high expectations for our students, and many of us would maintain that we do, in fact, maintain them. If we reflect on the complex picture painted by the research, however, it becomes clear that many factors mediate our expectations for our learners. For example, when we believe talent is innate, having to work hard becomes a sign that one lacks talent. If science, by its very nature, requires hard work, it is but a short leap to assume that there are few who can and should pursue it. Although Triad focused specifically on gender, we strove to set aside the stereotypes common in our culture and approach one another and our students as individuals. That said, none of us could claim to be completely free of the assumptions about "the other" that have resulted in our cultural divisions on the basis of gender, race, religion, class, sexual orientation, and disability even as we worked in schools to dismantle those same assumptions. Finally, out of the best of intentions we teachers guide and advise our students to pursue what we hope will be fulfilling and successful paths. Our beliefs about learning and the characteristics of our learners manifest themselves in our advice.

Our first step in Triad was to acknowledge that the issues were complex and rooted in a deep and often troubling history. We also honored the commitment of each member of our community to making change. Part of that honoring involved working together to define some concrete behaviors we could employ in our classrooms. We invite you to explore these issues and add your own ideas to the list.

- Talk with your students about the relationship between effort and intelligence, and let them know they have the power to improve their work. Maintaining high expectations means that all members of a learning community, teachers and students alike, are capable of developing their skills and aptitudes through

the application of effort. This does not mean that everyone will leave a set of experiences with the same understanding or that everyone will be at the same place at any given time. It does mean that, at the end of the day, every individual's understanding and skills should be deeper and more richly connected to other things they know.
- Identify the confusing concepts in your curriculum, and structure lessons in a way that gives your students the opportunity to work together as they grapple with them. Rather than allowing your students to rely on you or turn quickly to you for solutions, answer their questions with questions designed to help them clarify and deepen their own thinking. Think in advance about the kinds of questions you might use.
- Provide opportunities for your students to practice. We all know from experience that doing something really well requires skills that can be honed only through practice. This is true for playing an instrument, planting a garden, or planning a scientific investigation.
- Involve all members of the classroom community in giving and receiving substantive critical feedback. Getting better at something requires that we know what we're doing well and what we need to work on. Providing specific feedback is a skill in its own right, and teachers and students alike benefit from practice giving and receiving it.

Each of these strategies is essentially about believing that we all can be good at solving scientific problems and are expected to work hard to do so. When we consistently communicate these beliefs through our actions in the classroom, we maintain high expectations for our students and help our students do so for themselves.

The More We Expected

In this year of Triad, I also learned that I can and should have high expectations of the girls. The more we expected them to understand and the more we pushed them, the more they gave. In our last club meeting, the girls received vials of flies and separated them by sex and phenotype (red eyes versus white eyes). After a short discussion about genes, chromosomes, and dominant and recessive traits, the girls played with cards representing chromosomes that we used to demonstrate what happens during meiosis and gamete fusion. I was pleasantly surprised to find that nearly all the groups of girls were able to use the cards effectively, a task I thought might be too abstract for a middle school group. By keeping our expectations high, the girls were able to achieve more. Although I had understood that in theory, I didn't realize how well it could work in practice until that club meeting.

Reflection Questions

- "I also learned that I can and should have high expectations of the girls." What do you think this scientist means?
- What does it look like in a classroom to have high expectations of all students?
- Research shows that teacher expectations are a predictor of student success. What kinds of factors influence your expectations of students? How have you become aware of any differing expectations that you have for different students?

Links

- Student Goals: No Longer the Same
- Student Goals: I Don't Even Know How to Use a Saw

CHAPTER 5: Teaching Goals

- Science Goals: You Can Lead a Horse To Water
- Teaching Goals: Can You Help Me?
- Teaching Goals: Answering Student Questions With Questions

Teachers learned that high expectations led to higher student achievement.

SECTION II: Exploring the Triad Framework

Talking in Questions

This vignette was recorded by an observer conducting a small-group interview with several middle school girls at the conclusion of an all-girls after-school science club.

Girl: Yeah, and having the scientists, they cared, and they talked to you. They didn't talk to you like they were the scientist and you were the kid. They talked to you like you guys were friends. They talked like, "How do you think this will work?" and things like that, and not like, "You should do this or you should do that." They talked to you in questions, so you can actually decide, which isn't that common in other science classes that I've had where they're like, "You have to do this" or "You have to do that, and you have to decide what you're changing in this experiment and what's going on." It's not like that with the scientists. It's like, "What do you see that's changing?" and "What do you want to do that will change it the most?" It's not like, "Well, you have to try all these things and the one that changes it the most is the one that you'll do." So it's different.

Reflection Questions

- What does this girl value about her interactions with the scientists in her science club?
- "Talking in questions is a strategy for maintaining high expectations for students in a science classroom." To what extent do you agree or disagree with this statement?
- What is the balance between raising questions and answering questions? How does one decide when it is time to give answers? How is this balance related to maintaining high expectations for all students?
- How do you use questioning as a teaching strategy in your classroom? What kinds of questions do you tend to ask?
- What kind of classroom culture might best encourage students to be confident in asking their own questions?

CHAPTER 5: Teaching Goals

Links

- Student Goals: No Longer the Same
- Science Goals: To Trust in Their Own Logic
- Science Goals: Walking Encyclopedia
- Science Goals: Answers Are Not the Goals
- Science Goals: The Balloon Droops
- Teaching Goals: Resurrecting Socrates
- Teaching Goals: Does This Bridge Look Better Than It Did the Last Time?

> Another valuable lesson I learned from Triad is the idea of keeping your hands in your pockets. It has always been my first instinct to show a child how to do something when they ask for help. After watching the staff at Triad implement this strategy during our activities, I see the value in it. By answering questions with questions and making the child come up with the answer, you are empowering the child. They leave with a sense of accomplishment and pride. I have definitely implemented this strategy in my own classroom. I may talk them through it, but I will never again reach for their pencil. I've found that after helping them in this manner, their most frequent response is "See, I am smart!"
>
> —*Triad Teacher*

Can You Help Me?

This vignette is from an observer's notes during a large, all-girls, middle school, after-school science club. There were 10 small-group work stations with two to four girls at each station. The students were working on the second part of a greenhouse-building activity, assembling the pieces of wood they cut during a previous meeting.

Most of the girls were either cutting pieces of plastic from large black trash bags, using tacks to attach the plastic to each piece of wood, or hammering the four covered sides together. The activity was fairly difficult for the girls. It required lots of team coordination—one person to align and hold edges, another to hammer, and a third to keep the whole thing from falling off the desks. The task was further complicated by the fact that the wood pieces did not have straight edges.

A few girls worked quite independently. Others were shouting out for assistance, "Can you help me?" The club sponsors were kept busy assisting everyone. There seemed to be a tack shortage and groups began to share supplies. In one group, a girl sat cross-legged on top of the desk resting her chin in her palm while holding one side of the greenhouse down with her rear-end, as her partners hammered on the second side. This set-up seemed pretty efficient, but not too safe, as her back was turned toward the talk and the action.

A bit later, one club sponsor intervened and the energy level of the group jumped up. There was lots of excited laughter, and they were all holding and hammering corners together. Another club sponsor said, "Excellent! Well done."

A pair of girls were working quite independently near the door and had already put together the four sides of their greenhouse. The other groups were not as far along, but were much more talkative as they worked—sharing hammering techniques and suggestions for removing wayward nails.

I talked with a twosome briefly and discovered that they were both seventh graders, from different classes, and both were in Triad for a second year. They told me, "We're done. We just need to put plants in and the plastic cover." They continued working,

CHAPTER 5: Teaching Goals

cutting a section from a large roll of clear, thick plastic and taping it to the cover of their greenhouse. Other groups nearby were beginning to shout out, "We're done!"

After the greenhouse-building activity ended, the adults debriefed as they cleaned up. I asked the team how they thought the activity went. They mentioned trying to pay attention to safety issues, such as hammering near someone's head. They also talked about struggling to "keep their hands in their pockets and not simply step in to do the work for girls." They had the perception that students were more frustrated than they normally were. The work was harder than it looked, particularly because the precut pieces of wood were crooked. They noted one student in particular who always says, "I need help," but who can actually do it herself, and prefers to work alone.

Reflection Questions

- How do you respond when students ask you for help during a science activity? What guides your response?
- What skills can students learn by struggling through a physical or conceptual difficulty without the help of an adult?
- If we as teachers maintain high expectations by not taking over and rescuing struggling students, then how do we also keep other students from doing this as well?
- When teaching a student-directed science lesson, what classroom-management strategies might you use with respect to student grouping, student roles within groups, safety, and materials management to facilitate positive and productive student interactions while still maintaining high expectations?

Links

- Student Goals: Safety Was a Concern
- Student Goals: I Don't Even Know How to Use a Saw
- Student Goals: Don't You Feel Powerful?
- Science Goals: Nobody Knows What's Inside
- Science Goals: The Balloon Droops
- Science Goals: To Trust in Their Own Logic

Answering Student Questions With Questions

Although Triad was a great opportunity for the girls to do a lot of learning, I think that I was the one who got the best education. Because I am a scientist, some of the most important lessons I have learned deal with teaching techniques and practices. My partner teacher was a terrific mentor on how to best lead a class discussion and get students to think more deeply and conceptually about a problem. She primarily led group discussions by asking questions and in turn answering student questions with questions. It may seem an obvious method, but it is extremely difficult in practice. The results are what are most impressive. In our lesson on testing the areas of school for germs—by swiping surfaces with Q-tips and transferring the bacteria to an agar plate—the girls participated in an in-depth discussion about how bacteria grow, how our bodies react to invading germs, how we can kill germs, and the effect of antibiotics on germs. These are conceptually difficult ideas for anyone and particularly girls this age; however, it was evident that with a well-implemented discussion, the girls could think in a very sophisticated way.

Reflection Questions

- What teaching strategies has this scientist found rewarding to learn? What has she found most surprising about working with these students?
- In what ways can answering students' questions with questions aid us as teachers in maintaining high expectations of students? How can it aid students in developing more sophisticated thinking skills?
- How do you feel when someone responds to your question with another question as opposed to an answer? Why do you think you feel this way?
- If a student responds in frustration when her question is met with another question, how might you deal with that situation?

CHAPTER 5: Teaching Goals

- "The higher my expectations are for students, the more I see them rise to the challenge." To what extent do you agree with this statement? To what extent do you disagree?

Links

- Student Goals: Where to Draw the Line
- Student Goals: We Have Reason To Believe
- Science Goals: Answers Are Not the Goals
- Teaching Goals: Keeping Your Hands in Your Pockets
- Teaching Goals: Back in the Classroom

Teachers learned to get students to think by answering questions with questions.

SECTION II: Exploring the Triad Framework

Theory Is Easy, Practice Is Difficult

We did two club sessions on the mystery box. We tried not to make it into a competition. The first group that had a working box demonstrated it to the class, and some were also demonstrated at Family Science Night, although most groups didn't finish. This was one of the most demanding clubs for me personally as I was trying to help a pair of girls—one eighth grader with (as I learned later) learning difficulties and a bright-but-shy sixth grader. It was a pairing that just didn't work. I kept to the principle of not giving away ideas and instead asked questions, but the magic didn't happen—an example of the theory being easy and the practice, very difficult. We discussed some ideas for how to get the box to work, a couple of which had some validity. I managed to drag out an idea for the first problem—how the clear water becomes colored—and they tested this. However, they couldn't come up with any ideas on the second problem (how the water then changed color upon pouring water into the box a second time). I couldn't tell them and couldn't work out a way to lead them to a solution with questions without telling them how to do it. This pair did not get anywhere near making a mystery box. I felt there should have been something I could have done—but I still don't know what.

The last club was a cow-eye dissection. Again, I seemed drawn to the girls with the biggest problems. However, this time some magic happened and, after I forced myself not to pierce the eye myself out of frustration, the two girls I was helping did manage to dissect the eye themselves. This situation has issues related to my enjoyment as a teacher. On the one hand, these girls were in need of time and encouragement. On the other hand, it was like pulling teeth, and, even when they achieved something, there was no real interest in the science. I am not the type of person who gets job satisfaction in these small steps. I felt it was something I had to do because I was the one with sufficient patience to help, but I don't really enjoy it and I would much rather be helping the bright eighth graders who ask question upon question and are naturally curious, but my time was taken up elsewhere. So how do I work this out?

CHAPTER 5: Teaching Goals

Reflection Questions

- What issues is this scientist struggling with in interacting with students? Does she maintain high expectations for all the students she describes, even during her struggles? If so, how?
- Have you had experiences similar to this? If so, how did you respond?
- What might you have done with the struggling mystery-box pair if they were in your classroom? With the struggling cow-eye dissection pair?
- How can experiencing frustration benefit students? How might you decide when to intervene and bring student frustration to an end?

Links

◯ Student Goals: Where to Draw the Line

◯ Science Goals: You Can Lead a Horse to Water

◯ Science Goals: Nobody Knows What's Inside

◯ Science Goals: A Daunting Task

Eliciting interest in the science involved in an activity can be "like pulling teeth," said one Triad participant.

Girls in Science

SECTION II: Exploring the Triad Framework

Science Is Not a Priority for These Students

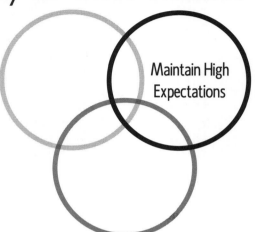

As we prepared for what I considered the most exciting club meeting, the heart dissection, I was sure more girls (especially African American students) would catch the fever and join. We sat in the science lab and discussed how to dissect. Many African American girls came to visit. They watched and became very interested in the process of cutting arteries and slicing heart valves. We continued to discuss and prepare in their presence. In between, I tried to encourage the girls to join and attend the club. They seemed excited, but, on the day of our actual meeting, very few showed up. This was disappointing. I understand that there are restrictions that apply to why so few African American students attend, such as commitments at home and lack of transportation, but my initial reaction is that science is just not a priority for whatever reason. How can I develop a desire in these students? What more can I do to spark an interest? These are some of the challenges that I faced.

Reflection Questions

- What is your response to this teacher's description of struggling to involve more students in the science club?
- Would your response to this account be different if you knew this teacher is an African American woman teaching in a school that has 11% African American and 44% Latino and Latina students? Why or why not?
- Do you agree with the author that student interest is the issue here? Why or why not? If not, what else might be impeding student participation?
- What relationships have you observed in your own classroom between race, class, culture, and gender and how these characteristics correlate with engagement in school science? What strategies have you used to engage those students who seem to be the least involved with school science?
- Have you encountered examples of differing expectations for student success in science based on students' personal characteristics? How have you responded?
- How would you approach constructing a classroom environment where there are equal opportunities for all students to reach a high level of success in science?

CHAPTER 5: Teaching Goals

Links

- Student Goals: After the Initial Eeewwww
- Science Goals: Turning to Nadya
- Science Goals: The Most Difficult
- Teaching Goals: Talking in Questions
- Teaching Goals: Accepting Stereotypes

SECTION II: Exploring the Triad Framework

Accepting Stereotypes

I always knew that equity was important in the classroom setting. I spent my middle school years in Texas where I was one of two Asian students in the entire school of more than 800 students and spent the majority of that time trying to battle the stereotypes that followed me. I was consistently classified as the "brain" by teachers and students alike. Most students assumed I would not be interested in friendships or social activities even though I was quite lonely. Their assumptions about me led me to conform to their stereotypes, and for a long time, I was quiet, shy, and unsure of myself socially.

During the Triad retreat, we did a mobile activity in which different groups started out with a different amount and variety of materials. This time, I was on the other side and had been given so many colors of paper, types of string, and different markers and crayons that it was literally impossible to use them all. Interestingly I did not notice that others had received less than my group had, nor did I recognize their resentment. Just like me in middle school, they did not speak out because they assumed that it was not OK to ask us to borrow our scissors or use some of our string. I learned that assumptions about social rules of behavior flow both ways. I initially attributed my own experiences in middle school purely to the teachers' and students' assumptions about me, but I was equally to blame for accepting those stereotypes and playing up to them by acting shy and quiet even though my natural personality is outgoing.

Reflection Questions

- What kinds of effects do you imagine the expectations described in the first paragraph had on this individual as a middle schooler?
- To what extent do you think it's reasonable to expect a middle schooler to resist such expectations? Explain.
- How are your expectations of your students colored by your perceptions of their ethnicity, race, or class?

CHAPTER 5: Teaching Goals

- Describe an experience in which your expectations of a student were significantly changed as you got to know him or her better.
- What kinds of active strategies might you use to curb the expectations that come with your cultural experience so you can encounter each of your students as individuals with unique sets of attributes?

Links

Student Goals: On a More Personal Level

Science Goals: Putting Sugar in Water

Teaching Goals: Science Is Not a Priority for These Students

Teaching Goals: At First I Was Hesitant

SECTION II: Exploring the Triad Framework

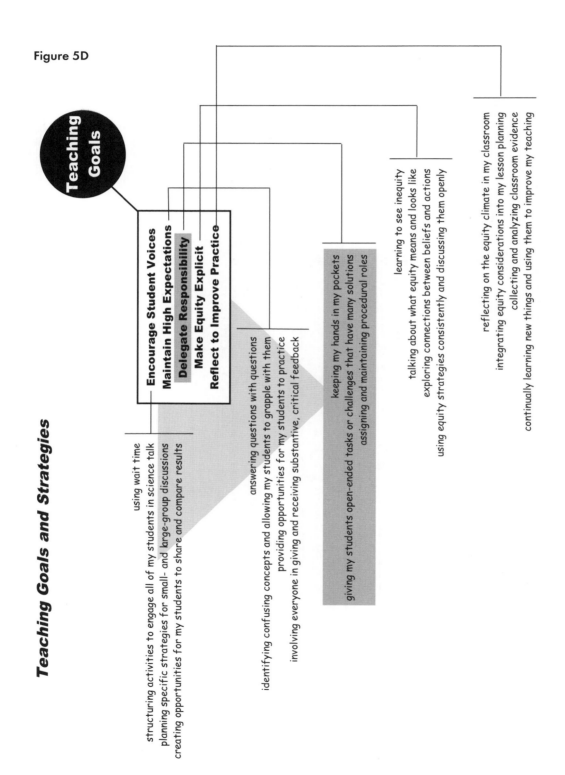

Figure 5D

Teaching Goals and Strategies

Teaching Goals

- **Encourage Student Voices**
  - using wait time
  - structuring activities to engage all of my students in science talk
  - planning specific strategies for small- and large-group discussions
  - creating opportunities for my students to share and compare results
- **Maintain High Expectations**
  - answering questions with questions
  - identifying confusing concepts and allowing my students to grapple with them
  - providing opportunities for my students to practice
  - involving everyone in giving and receiving substantive, critical feedback
- **Delegate Responsibility**
  - keeping my hands in my pockets
  - giving my students open-ended tasks or challenges that have many solutions
  - assigning and maintaining procedural roles
- **Make Equity Explicit**
  - learning to see inequity
  - talking about what equity means and looks like
  - exploring connections between beliefs and actions
  - using equity strategies consistently and discussing them openly
- **Reflect to Improve Practice**
  - reflecting on the equity climate in my classroom
  - integrating equity considerations into my lesson planning
  - collecting and analyzing classroom evidence
  - continually learning new things and using them to improve my teaching

Essay: Delegate Responsibility

Science is a discipline in which our understanding and explanations are constantly evolving. Doing science requires one to question, to approach results with skepticism, to perturb, to control, to analyze, to wonder, to persist, and to convince. As one Science and Health Education Partnership scientist put it, "Science is disciplined rebellion." If this is the nature of scientific activity, then where does authority rest? Ultimate authority resides within the natural world of phenomena and the evidence it provides. In the human world, scientific authority is distributed across geography, time, cultures, institutions, and individuals. Science is inherently social; if it is to be used for the good, it requires that everyone who participates in the production and application of scientific knowledge do so with a spirit of independent and critical inquiry. This distributed authority has implications for how we teach science. To engage in the making of their own scientific knowledge, our students must learn to make decisions—ranging from interpreting a step in a protocol to deciding how to design experiments in order to answer a question—and we must give them the authority to do so.

Sociologists Elizabeth Cohen, Rachel Lotan, and their colleagues have studied the relationship between delegating authority to students and optimizing the engagement and verbal participation of the learner. Their group-work approach, called *complex instruction* (CI), is based on the sociology of groups and designed to promote the engagement of all students. Central to CI are lessons designed to foster development of higher-order thinking skills through the use of multiple abilities and group-work activities organized around a central concept. In addition to being open-ended and requiring student interdependence to solve problems, the tasks require a wide array of intellectual abilities so that students from diverse backgrounds and different levels of academic proficiency can make meaningful contributions to the group task (Gardner 1983). Educators train students to use cooperative norms and specific roles to manage their own groups. This delegation of authority to students supports their persistence and problem solving and frees the educator to observe groups carefully, to provide specific feedback, and to treat status problems that cause unequal participation.

The treatment of status issues—recognizing status problems and conferring status on learners through recognition of intellectual contributions—is an essential aspect of CI and helps ensure equal access to learning. Status and status treatments are of particular interest for gender equitable teaching. The more students talk on task while manipulating relevant materials, the more they learn (Cohen and Lotan 1997), and

SECTION II: Exploring the Triad Framework

status influences these behaviors. There are different kinds of status in the classroom: expert status, academic status (being very good in the subject of study), peer status (being popular and attractive), and societal status (Cohen 1994). These different kinds of status generalize: A student who is good at reading is likely, as a result, to have high status—which then can transfer to an area in which the individual does not have expertise but to whom others defer anyway. This is because there are general expectations for competence that are attached to high and low status: On any important new task, high-status individuals are generally expected to do well whereas low-status individuals are generally expected to do poorly. This results in expectations for competence and incompetence for high- and low-status learners respectively and sets the stage for self-fulfilling prophecy. High-status learners engage in more talk and manipulate more materials, thus learn more; low-status learners talk less, have less opportunity to handle materials, and consequently learn less.

So what does this look like in a science classroom? Science, with its orientation around process and inquiry, is a natural subject for delegating authority to students. Open-ended investigations involve students in making decisions, solving problems, manipulating equipment, analyzing results, communicating findings, defending positions with evidence, examining the work of other scientists, and learning from their mistakes. When students are designing their own investigations and engaging with one another as members of a community of scientists, teachers have increased opportunity to observe, to pinpoint interaction and engagement, and to stimulate thought. As stated in the *National Science Education Standards*, "The teacher's role in group interactions is to listen, encourage broad participation, judge how to guide discussion—determining ideas to follow, ideas to question, information to provide, and connections to make. In the hands of a skilled teacher, students recognize the expertise that different members of the group bring to each endeavor" (NRC 1996, p. 36).

In Triad we employed a variety of strategies as we worked to delegate more authority to our students. Some were fairly simple and others complex. We found that these strategies gave us places to begin and subtle teaching methods we could work to develop over time. We hope you find this to be true, as well, and that you add a few of your own.

- Keep your hands in your pockets. Literally or figuratively, keeping your hands in your pockets helps you refrain from jumping in and solving problems for students. Whether they are struggling with how to use a scientific tool or how to approach data collection in an investigation, they learn more when they have opportunities to help one another through their struggles. They may need verbal instructions on how to

CHAPTER 5: Teaching Goals

focus the microscope, but the act of adjusting the focus itself is something they can and will learn to do. Likewise, a group may need some feedback on the design of an investigation, but that feedback can come through questions that help them assess and refine their design rather than replacing it with a design from the teacher.
- Give students open-ended tasks or challenges that have a variety of solutions. Such tasks not only engage students in making decisions, but also tend to reveal multiple abilities that can be publicly acknowledged for low-status students. Open-ended investigations require thoughtful planning and substantial preparation on the part of a teacher, so it may not be possible all of the time. But, with every science investigation we do, we can ask ourselves how we might provide increased opportunities for our students to make decisions, solve problems, and learn from their mistakes.
- Assign and maintain procedural roles. Well-structured group work utilizes differentiated procedural roles for group members—such as facilitator, recorder, materials manager, and timekeeper—that correspond to the group task. Over time, every student has many opportunities to perform all of these roles. Such roles serve to delegate the authority of accomplishing the task to the group, making students responsible for their own learning and for their group mates' learning. It is important to remember that these roles are procedural rather than intellectual. In other words, they help the group organize to get the task done, but every member of the group is intellectually responsible for the work regardless of their assigned role.

Effective use of group work, like that employed in complex instruction, is not simple and takes learning on the part of teacher and students alike. The key concept of delegating authority, however, can be approached in a variety of ways and is essential to providing students with opportunities to develop and practice the kinds of decision-making skills that are characteristic of the practice of science.

SECTION II: Exploring the Triad Framework

Keeping Your Hands in Your Pockets

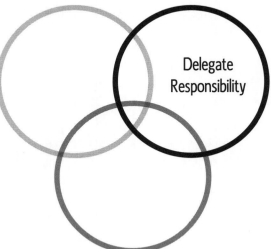

In my classes and in the club I always tried to remind myself of the goal, "Keep your hands in your pockets." It was easier to stick to this in the club because we had more time and because we had more adults. In my classes this was always a challenge. The time seems to run short when students don't listen and then don't know what to do. Sometimes if I don't do some of the steps for them, they fail, and we don't have another class period to try it again.

As I started to see the value of keeping my hands in my pockets, I planned some lessons so that students would be given time to explore materials without my having to show them what to do. I used the electromagnet lesson that we used in last year's club in my eighth-grade science classes. Students were given a nail, a battery, some paperclips, and a wire. Without showing them how to build the magnet, I gave them about 20 minutes to see if they could figure it out. I used this same format for lighting a lightbulb and inventing a magnet machine. I found particularly in the English-as-a-second-language class that students were engaged because they were given time to explore and figure it out on their own.

Reflection Questions

- What does the phrase *delegating responsibility to students* mean to you?
- In what ways is keeping your hands in your pockets related to delegating responsibility to students for their learning?
- How much student frustration in doing a science activity is too much frustration? How would you decide when to intervene and take your hands out of your pockets?
- When intervening, how would you choose from each of the following responses and why: 1) asking the students a question, 2) telling them how to do something, 3) showing them how to do something with a separate set of materials, and 4) doing something for them with their own materials?

CHAPTER 5: Teaching Goals

- Research suggests that adults are more likely to tell a boy how to do something, but actually do it for a girl, a phenomenon sometimes referred to as *rescuing*. To what extent have you examined your own responses to girls' and boys' requests for help during science lessons? What do you predict you might find?

Links

- Student Goals: Don't You Feel Powerful?

- Student Goals: After the Initial Eeewwww

- Student Goals: Where to Draw the Line

- Science Goals: To Help and Teach Each Other

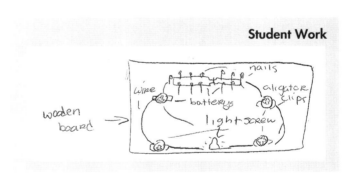

Student Work

SECTION II: Exploring the Triad Framework

Does This Bridge Look Better Than It Did Last Time?

This vignette is from an observer's notes of a bridge-building activity during a coed after-school science club for seventh and eighth graders.

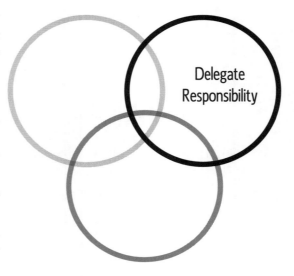

Teacher: One thing we're going to do differently this time is that if people don't like their designs they can redo it, but they have to use the same 30 pins and the same straws. There's nothing wrong with redoing your design—that's what experiments are all about. So why don't we break into our groups? Discuss with your partner how you're going to design and build your bridge.

[I heard the following questions being asked of the students by the teachers and scientists as they circulate among the bridge-building groups.]

- What's happening to your bridge?
- What can you do to make it stronger?
- What could you do with those straws?
- The test space is open, do you want to try it out again?
- What did it look like last time?
- Let's not think of it as a competition. Ask yourself "Does this bridge look better than it did last time?"
- Is there any way that you can reinforce that?

CHAPTER 5: Teaching Goals

Reflection Questions

- What kinds of questions are being asked? What additional kinds of questions would you be asking students?
- During this activity, what do you think the teachers and scientists are doing? What might they purposely not be doing?
- How are the teachers and scientists delegating responsibility to students? In what other ways can you delegate responsibility to students in a science classroom?

Links

 Student Goals: So That All the Bridges Fall

 Science Goals: To Help and Teach Each Other

 Teaching Goals: Talking in Questions

> Today you are a bridge engineer! Your job is to build a bridge that can do two things. First, your bridge must span a 1 m gap. Second, your bridge must be able to hold the weight of at least 3 toy cars. Just like real bridge engineers, you have limited materials to work with. Each team will have 50 straws and 50 pins with which to build their bridge. There are two testing stations in the room where you can take your bridge and test it with the toy cars, and you can go to the testing stations anytime you want while building your bridge. Like all engineers, your first design probably won't work. So, keep trying and keep improving your design!
>
> —*Triad Scientist*

I Could See How Much I Learned

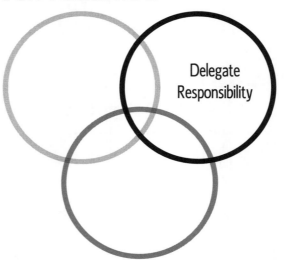

In addition to giving the girls experience in being scientists, we realized that it was important to let them gain experience in teaching. We structured two activities to allow girls to teach other girls how to do something after they learned it themselves. I realized how important this was on Family Science Night, after one girl in our club excitedly told me, "It was really cool to show my sister how to do the experiment. I could see how I must have been before I learned how to do it. I could see how much I learned. Is that how you feel when you teach us? It's really nice." I couldn't have said it better myself!

Reflection Questions

- How can engaging students in peer teaching and peer learning delegate to them responsibility for their own learning?
- "You never really understand something until you have to teach it." What about this statement do you agree with? Disagree with? How does your own understanding of a topic change when you have to teach it to others?
- In what ways can students' experience teaching a science concept to other students change their own understanding of the concept? How might it change their perceptions of themselves in science and as learners?
- What kinds of tasks or topics within your science curriculum lend themselves well to having students do the teaching?
- When students are in the position of teaching others—with, for instance, class presentations, jigsaw discussions, scientific posters—how do we build in opportunities for feedback and corrective input from the teacher without undermining students?

CHAPTER 5: Teaching Goals

Links

- Science Goals: The Strength of the Group
- Teaching Goals: See What Happens
- Teaching Goals: I Have to Introduce Triad

Girls in Science   217

A Different Role

A number of experiences gained at the fall retreat strengthened my commitment to the Triad goals and helped me lead club meetings later in the year. When I first joined Triad I was aware of gender inequity, but I was unaware how prevalent such inequity has been within the scientific community. I began to appreciate the strong feelings on the subject amongst women participants at the retreat when I learned how so many of them had to struggle to overcome such inequality during their careers.

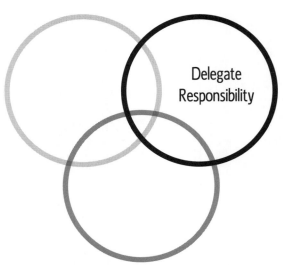

The workshops at the retreat proved to be particularly useful. Assigning each member of a team a different role for each experiment demonstrated how effective this technique was in preventing a single member of a team from either taking over the central role or from being relegated to little or no role at all. Later in the year in our Triad clubs, my team members and I made a conscious effort to similarly ensure equal, fair participation of each club member in our experiments.

Reflection Questions

- When delegating responsibility to groups of students, what structures must be in place to avoid any one student taking over the group?
- Recall a time when you experienced or witnessed gender inequity in a science classroom? During that situation, were roles assigned or assumed?
- What classroom culture and structures must be in place for assigned roles to be helpful to students as they work in groups?
- Research has suggested that the size of a group of students working together should be dictated by the complexity of the task, not the availability of materials, even though this is not the reality of most classrooms. How do you decide on the size of student groups during science investigations? How do you choose who will work together? To what extent have you experienced single-sex groupings in your classroom? Why or why not?
  - "I strongly dislike working in groups." To what extent do you agree with this statement? To what extent do you disagree? Why?

CHAPTER 5: Teaching Goals

Links

- Student Goals: She Wanted to Do It Herself
- Science Goals: Turning to Nadya
- Science Goals: The Strength of the Group
- Teaching Goals: Personal Development
- Teaching Goals: Anyone But the Boy

Possible Roles for Students Working in Groups

Facilitator: Makes sure that everyone gets the help she needs and is responsible for seeking answers to the group's questions. The questions are only posed to the teacher if no one in the group can help.

Materials Manager: Gets all the necessary materials as instructed by the teacher. Makes sure that materials are being used properly and safely.

Timekeeper: Works to ensure that everyone in the group is on task and that the group is adhering to the agenda.

Recorder: Records the group's work—this might include the procedure, materials, data, results, and conclusions.

SECTION II: Exploring the Triad Framework

No One Felt Uninvolved

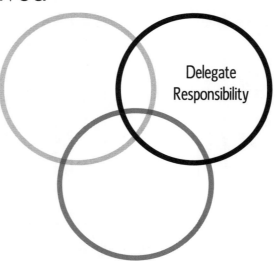

Learning skills—such as being explicit about equity—and implementing those skills in the classroom or workplace are two very different things. Although I feel I have learned the philosophies, skills, and techniques presented and demonstrated by the Triad staff, I still have a long way to go in learning how to use them. Only toward the end of the year did I find myself really understanding when and how to use what I had learned. I don't feel I ever really used the skills effectively in my club. However, they have proved useful in my work setting. Here is one example of how I have successfully used role assignment in my laboratory setting.

I work on a developmental biology project involving zebra fish as a model organism with two other research students. Right now the project is in the preparation phase. This entails a lot of fish maintenance: crossing fish weekly, separating and counting the eggs, feeding and raising these offspring, and transferring them to new tanks when they are big enough. One of the postdocs decided she wanted to be in charge of the project for a little while until the fish gave better numbers of eggs. Unfortunately, this ended up being a situation in which the other postdoc and I were left completely out of the loop for three weeks straight. Feeling frustrated at my lack of involvement in the project, I called a meeting of the three of us to discuss the progress of the project so far.

Using what I had learned from the Triad workshops, I carefully planned for the meeting. I drew up an outline of topics to be discussed and came with a list of all the different jobs that the project entails. Acting as mediator, I would bring up a topic of discussion and ask the others for any input they had. After giving my colleagues time to express their thoughts, I would add mine. At the end of the meeting, we talked together to assign roles for each of us on a weekly basis. During the first week, I would cross the fish and collect and count the eggs. One of my colleagues would be in charge of feeding and raising the small fish, and the other colleague would be responsible for moving any fish that were big enough to new tanks. During the next weeks we would rotate jobs. Having all of us know our roles for the week made us work better as a team and made it so no one felt uninvolved or uninformed.

Reflection Questions

- What does this scientist's story have to do with "delegating responsibility to students?"
- Have you ever worked on a group project that used explicitly assigned roles? What worked well? What didn't work so well?
- When you have a role in a group, does that mean you no longer have intellectual responsibility for the whole project? Why or why not?
- To what extent have you used procedural roles—such as timekeeper, recorder, reporter, facilitator, materials manager, and equity monitor—to help structure group work among your students during scientific investigations? What are the benefits? What are the challenges? How do you talk with students about their procedural versus intellectual roles in a group?
- How might delegating responsibility to students during science learning transfer into other aspects of their life at school and at home?

Links

 Student Goals: On a More Personal Level

Student Goals: The UV Bulb Can Be Changed by the User

Science Goals: Turning to Nadya

SECTION II: Exploring the Triad Framework

Figure 5E

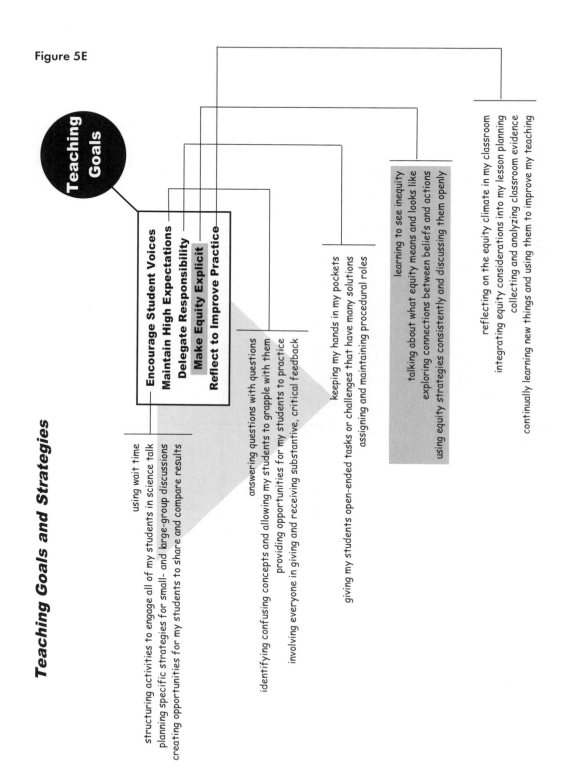

Teaching Goals and Strategies

CHAPTER 5: Teaching Goals

Essay: Make Equity Explicit

Do an internet search for famous scientists. Pick a list and take a look at the names. Whether you choose a list that focuses on scientists in general or scientists within a particular field, the names you find will be almost exclusively male. Go to a library and pick up a book on the history of science. In almost every case the story will begin in Ancient Greece and move to Western Europe with males as the main characters. Do a quick search for posters of famous scientists. The picture improves a little. You will find a women-of-science series, a great-black-innovators series, and a few independent posters of Japanese physicists. These swim in a sea of images of Einstein, Newton, Darwin, Freud, Pasteur, and Galileo (just to name a few). So what do such quick surveys tell us? Nothing that most of us did not already know—science, as it has been recorded by posterity, has been a field in which men were the primary actors. This reflects not only the easier access men have had to science but also the story that has been told about science. So, although you can find lists of important women in science compiled by feminist scholars, reading through these lists makes it clear that their history has not been well recorded or passed on. This also holds true for the contributions of people of color. When the perspectives of so many of the world's people are excluded from scientific discourse, what is lost? What are the questions that don't get asked, the lines of inquiry that don't get followed, the ways of coming to understand the natural world that don't get honored? Given the many challenges we humans face, it is vital that the scientific community begins to talk explicitly about these issues.

There is a large body of research on how students and teachers view scientists. Often referred to as "the draw-a-scientist literature"—in reference to a particular instrument that has been used to assess these views—it paints a disturbing picture. Since 1957, when the first such study was done (Mead and Metraux 1957), a fairly consistent image of who the scientist is and what the scientist does emerges from both student and adult data. Time and time again, the scientist that appears is a white, male, often bald, bearded, glasses-and-lab-coat-wearing chemist. This is, of course, not true for every individual surveyed, but elements of the stereotype are persistent across gender, culture, geography, ethnicity, and time. Researchers have also consistently found a strong relationship between stereotypical views of scientists and student tendencies to enroll in science courses and pursue science as a career. The less stereotypical the view, the more likely students are to participate. Happily, the research also shows that the stereotype can be changed. Explicit attention to gender and racial equity and helping students to see scientists as regular people, all seem to have positive effects (Finson 2002). When students are able to see scientists as being not so different from themselves, they may also be able to envision themselves as scientists.

SECTION II: Exploring the Triad Framework

Students and teachers alike walk through the school doors knowing the world is not exactly an equitable place. Whether the inequities arise from the social and economic structures surrounding gender, race, or class, it is clear that students enter the classroom with differential amounts of privilege and disadvantage (for a discussion of privilege, see McIntosh 1990). As teachers we can do our best to create equitable learning environments so all of our students are learning successfully. Structuring activities so that all students have the opportunity to work with materials, building in time so that all students are able to talk about their ideas, and giving students the freedom to wrestle through challenging tasks may create a space in which students can learn a great deal about science and how it is done. It does not necessarily mean that they will understand why we choose to structure things in the way we do or how they might make choices of their own that would contribute to an environment in which all have access to learning. Developing an understanding of what equity means and why it is important requires that we be explicit with our students about the history, social structures, and resulting attitudes that have led to inequities within the field of science and the science classroom. The word *explicit* comes from a root that means to unfold or unroll. The challenge for us as teachers is to figure out how to "unfold" our intentions and related actions and to help students do so for themselves and with one another as well. We must help them to understand what equity does and does not look like, and support them as they gain skills in working together in equitable ways.

In most classrooms and, indeed in most of our daily lives, equity is not explicitly addressed. Talking explicitly about equity makes many of us uncomfortable. We fear that we will offend. We fear that we will reinforce negative stereotypes. We question our ability to facilitate the discussion or to manage conflict that arises. By avoiding such conversations, however, we help to perpetuate deep cultural assumptions about the relationships between individual effort and success—that we each earn the privileges we are granted or denied by society. Equity must become a lens through which a classroom is seen, and instances of equity and inequity identified, acknowledged, and grappled with. If students are to learn to value participation for themselves and all of their fellow learners, they need opportunities to talk about what participation means, to learn to recognize ways individuals avoid or are denied participation, and practice strategies for increasing the meaningful participation of all members of the community. They must also be able to see themselves reflected in the role models, histories, and topics of research that we include in our science curricula. The strategies we used in Triad were designed to help us all get a start:

CHAPTER 5: Teaching Goals

- Learn to see inequity. Teacher awareness is an important component of making changes in the classroom. When we are unaware of the inequities that exist and how they came to be, we have little hope of creating truly equitable learning environments. Educate yourself about how systems of privilege are constructed and maintained. Read, talk with others who are committed to making change, get to know people whose experience with equity in our culture has been different from your own. Gloria Ladson-Billings (2006) describes this as developing sociopolitical consciousness and includes both local issues of the community and larger issues that affect equity in her description.
- Talk about it. Much of the literature on racial equity emphasizes the importance of conversation (Landesman and Lewis 2006; Bolgatz 2005; Singleton and Linton 2005; Pollock 2004; Kivel 2002). Whether we're talking with adult colleagues or our students, these conversations take some courage. On the other hand, when we commit to talking with one another, we have the power to begin robbing sexism, racism, and classism of their power.
- Explore the connections between your beliefs and actions. Living in a culture in which it is uncommon to discuss equity openly, it is difficult to identify when our actions are in conflict with the beliefs we profess to hold. Find colleagues who are willing to explore this with you. Observe one another in the classroom, share teaching dilemmas, and take a close look at your curriculum and assessments. Provide one another with the supportive but critical feedback that can help you all get better at being the equitable teachers you want to be.
- Use equity strategies consistently and discuss them openly with your students. Help students understand why it's important for everyone in the classroom to have opportunities to work with the materials, contribute their ideas, and critique the products of the community. Teach them to identify when inequities occur in the classroom, and give them strategies for making change.
- Turn the lens of equity on the curriculum itself. Examine what is included and not included in the official curriculum—the official version of science—and help students to master this knowledge and to see what has been omitted. Often what has been omitted is information that may be particularly relevant to students (Ladson-Billings 2006; Calabrese Barton et al. 2003).

These strategies are general. The specifics will arise from the context in which you work. What's most important is taking some first steps. It does not mean that you have to have all the answers. In fact, it may be more useful to just admit that we don't and to join with one another and with our students in talking openly about our experiences as we work to construct classrooms and a larger society in which we are equitable with one another.

CHAPTER 5: Teaching Goals

To Cunningly Mediate Equity

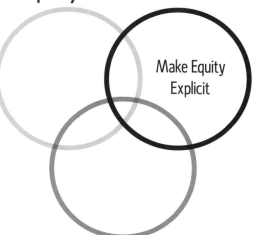

At first I thought it was up to me as a teacher to cunningly mediate equity within a group of students without them realizing that it was happening. Now I see this is not the approach I want to take. Equity can be achieved only through the awareness and understanding of everyone involved. I might start a lesson explaining that we are going to work in groups of four and that one goal of the activity is to make sure everyone gets a chance to add a gear to the machine we are building. I might even recruit the help of those students who know they are much more hands on than others by saying to the class, "Try to be aware of how much you are handling materials compared to others in your group. If you think you are doing most of the work, pass the materials to someone who hasn't had as much opportunity to build."

Reflection Questions

- What change has occurred in this teacher's thinking about how to approach equity issues in her classroom?
- What inequities are most evident to you in your classroom? In your school? In your community?
- How explicit are you with your students about equity issues in the classroom? With respect to gender? Race? Other personal characteristics?
- How can one prepare students to work equitably in groups? Be aware of and regulate their own actions within a group?

Links

- Student Goals: Watch Me!
- Student Goals: I Didn't Want to Produce the Same Fears
- Science Goals: The Strength of the Group
- Science Goals: I Learned How a Lava Lamp Works

SECTION II: Exploring the Triad Framework

The teacher of these students realized that her class also needed to know that equity was a goal of the lessons.

Like Dad

This vignette was recorded by an observer during an interview with an elementary school teacher. This teacher was a leader for a new, after-school science club for girls in the fourth and fifth grades. The teacher is responding to the question, "How did the first two clubs go?"

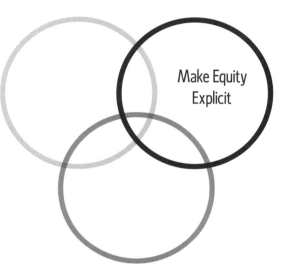

At the first meeting, we did a lot of talking about why this is a club for girls. And that—I wanted to tell you—was really a good discussion because we asked them. They did a pair-share in which they talked with a partner, and then they shared their ideas with the whole group. And they all came up with everything we could have said.

One girl said, "Well, when girls get to middle school they just turn into teenyboppers. They just talk about their hair and clothes and boys." And then somebody else said, "Oh, because boys are really noisy and they grab things." But then, and I was really glad we talked about this, another girl said, "So does that mean your dad is really noisy, and your dad grabs things? 'Cause you're talking about your dad!" and she laughed.

I was really glad that she said that, and I said, "I have students that are boys, and I love them. I think they're great. And I like my dad. I like my brother." I said, "This isn't about not liking boys or being better than boys." I explained that it was about making things more fair.

And also, another good point that came up, one of the students said after we talked, she just said, "I don't understand why everybody keeps saying that girls are not interested in school, because, well, at least from second to fifth grade," she said very specifically, "girls are the ones who are always quiet, and girls are the ones who are always doing their work and are interested in schools, and boys are always screwing around and playing. So, I don't know why people say that girls aren't interested in school."

SECTION II: Exploring the Triad Framework

Reflection Questions

- What ideas about girls, boys, and gender equity do these girls raise in this discussion?
- Where might a discussion like this have gone with your students? How might this discussion be different in mixed-gender versus single-sex settings? With older students? With adults?
- What are the benefits and challenges of being explicit about gender equity with students? What specific structures or activities could you use in a club or classroom setting to prompt a discussion of equity issues among students?
- What have been your own experiences as a learner or as a teacher in single-sex settings? In all girls settings? In all boys settings?

Links

 Student Goals: Watch Me!

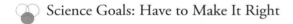 Science Goals: Have to Make It Right

Teaching Goals: Anyone But the Boy

CHAPTER 5: Teaching Goals

Talking About Equity

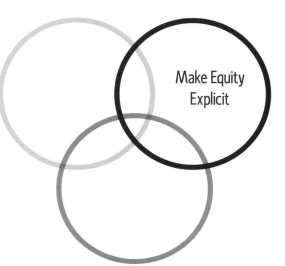

This vignette is from an observer's field notes of a group discussion on equity that involved teachers and scientists with at least one year of experience in Triad.

A Triad staff member summarized the questions we were to focus on: Do Triad clubs have explicit discussions about equity, and, if so, were they useful? How do we evaluate equity discussions we have with teachers, scientists, or students in Triad?

Teacher 1: I think it's hard. It requires a lot of planning and forethought. I've tried to open it up to the girls by saying, "Why do you think we have an all-girls science club?" and discuss some of their ideas. I'd like to address some of their questions. They might be interested in hearing some of the statistics. I think initially I was reluctant to have this conversation… because I didn't want them as fourth and fifth graders to have this expectation that girls don't do as well as boys. I think a lot of them had heard that their sisters in middle school were really into clothes and not as into school.

Scientist 1: There's always that fear that talking about stereotypes reinforces them. But basically ask them whether they think or how they think girls and boys are different in science. I'm not saying that they are, but there is this perception that they are.

Scientist 2: We need to create an environment in which you can have an equitable group discussion. Just yesterday, it was so difficult to call on the quieter girls. I asked a straightforward question and it didn't matter. The quieter girls did not respond. Maybe I'd stand there all day. There is the philosophy about talking about equity, but it's a totally different ball game when you are in there.

Scientist 3: Obviously discussing equity has its benefits. I think [the] point about time is a good one. If students are aware when they come into the club that this is about science and we're discussing equity, they might think, "So where's the science?"

Girls in Science

SECTION II: Exploring the Triad Framework

Reflection Questions

- Which of the scientists' statements do you agree with the most? Which of these statements do you disagree with?
- How can you talk explicitly about inequity with students without perpetuating stereotypes or inadvertently making them feel as if they are somehow deficient in science?
- How comfortable are you in being explicit about equity with your students? How do you think you could increase your comfort level? In what ways might you take small steps toward sharing with your students why you use certain teaching strategies to make your science classroom more fair and equitable?

Links

- Student Goals: I Didn't Want to Produce the Same Fears
- Student Goals: Watch Me!
- Science Goals: The Real Thing
- Teaching Goals: The Quieter Girls

> In the middle of the year, when one of the girls asked why we never let them work with their friends, we realized that we really could—and should—be explicit about equity. This led to an impromptu skit (by us) about cliques and about what it feels like to exclude or include people. The girls seemed receptive, so we decided to design skits for several later clubs. We demonstrated teaching each other skills—as opposed to doing someone else's work for them—and we demonstrated sharing materials. While these skits certainly didn't create perfect equity in the club, the girls did think about our messages. Immediately following an equity skit, I was asking the students at my table, one by one, to measure different boards. The girls noticed who had been left out and pointed it out to me, making sure that I was treating all of them fairly. Soon after that, we began sawing. We didn't have enough saws, but most of the girls shared without complaint, helping each other when they were waiting, and turning the tools over to other girls even before they were completely done.
>
> —*Triad Scientist*

Way Beyond Our Expectations

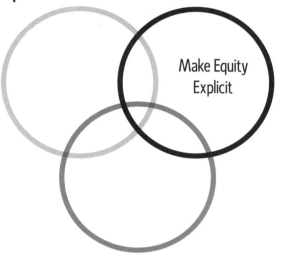

Our students' contributions were often impressively astute. In one meeting, we did the mobiles activity for which different groups were supplied with vastly different resources. We brought them together for the last half of the meeting to discuss how the different groups had carried out their task and how they had felt upon realizing that the room was rife with inequity. We had hoped they would be able to connect the activity to other experiences in school and in their lives in which differences in race, gender, and class make for inequity in achievement. In fact, the students went way beyond our expectations. Without our mentioning it, one girl immediately made a connection to the school science fair, where children with more family resources had a significant advantage over students with fewer resources or less parental guidance. These differences were reflected in their grades, and the students were vocal about their frustration. Another student rose beyond any of our expectations, describing eloquently his feeling of injustice relative to fellow classmates who, on account of the professions or financial stability of their family, received far higher marks even when their effort or ideas were far less substantial. He even defended himself, saying, "I'm not a communist or anything, but …"—a recognition of the political issues at stake that I had not expected from a seventh grader. In general, giving the students an opportunity to talk to each other and to the group freely was a slow but ultimately rewarding path.

Reflection Questions

- How were this scientist's expectations of the students in the club surpassed?
- What kinds of discussions about equity in both society and school have you explicitly had with either individual students or groups of students?
- What were these students' views on science fairs? What is your view? Your own students' views?
- Why do you think the student who talked about privilege and achievement felt compelled to say, "I'm not a communist or anything, but …"? What cultural assumptions, structures, and practices are being challenged by his observations?

SECTION II: Exploring the Triad Framework

- Recall a time when your expectations of a student have been surpassed? What effect did this have on your approach to teaching that student in the future?
- How do you interpret the scientist's last sentence? When have you had similar experiences?

Links

- Student Goals: After the Initial Eeewwww
- Student Goals: I Didn't Want to Produce the Same Fears
- Science Goals: Not What We Had Planned
- Teaching Goals: The More We Expected
- Teaching Goals: At First I Was Hesitant

Stop in My Tracks

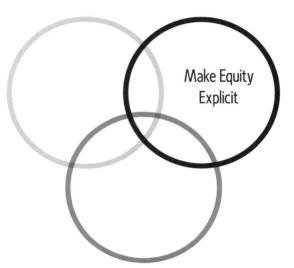

One of the main challenges I faced this year was how to address issues of equity directly with the girls. We planned a lesson, based on the mobile lesson, that was designed to bring up equity issues. Our lesson involved a tower-building competition, followed by small-group discussions on equity, and then a class sharing period. All of the teachers and scientist sponsors of the club were surprised that the girls were not very aware of the inequitable distribution of materials during the class. In addition, when we tried to get the girls to talk about equity in a more general way, we realized that they had not thought about these issues much and did not share many of our opinions and concerns about equity. The following week, when we surveyed the girls, they did seem more aware of the issues, but we were not sure how much they had actually thought about them or how much they were telling us what we wanted to hear. We brought up this issue at a Triad professional-development meeting, and people had many interesting comments on the topic. One comment, however, really made me stop in my tracks and rethink the whole lesson and approach to equity. This comment challenged whether we should really be forcing the topic of equity on students who don't seem to have any problems with equity issues. Is it beneficial to educate girls about equity if they have no concerns about equity themselves? On one hand, I felt like girls who are very privileged should be encouraged to understand their privilege so they would be more sensitive to others. However, our girls were not spoiled. From the surveys we collected from the girls, one of the main pieces of feedback we got from our lesson was the statement that "We learned that life is unfair." I have begun to question whether this is something that I want to be teaching to sixth-to-eighth graders.

Reflection Questions

- What was this scientist's experience with this one attempt at being explicit about equity?
- Do you believe middle school students are aware that life is unfair? Share an example of a conversation you've had or overheard that is evidence for your opinion.

SECTION II: Exploring the Triad Framework

- Have you been explicit about equity with your students? Why or why not?
- Do you believe it would be beneficial to be explicit with your students about equity? Why or why not?
- How can the diversity, or lack thereof, of a group of students—in terms of gender, race, culture, ethnicity, and other personal characteristics—influence their discussions about equity?
- Is being spoiled the same as being privileged? Why or why not? What does the term *privilege* mean to you? In what ways are you privileged? Not privileged?

Links

- Student Goals: I Didn't Want to Produce the Same Fears
- Student Goals: On a More Personal Level
- Teaching Goals: My Own Tendency
- Teaching Goals: At First I Was Hesitant

Building Mobiles Discussion Prompts

Share your experiences in building the mobile with people from other groups:

- How did you work with your partner to build the mobile?
- What did you notice about the materials that other groups had?
- If you were aware of it, how did it feel to have different materials than other groups?

CHAPTER 5: Teaching Goals

Anyone But the Boy

During an interview, a scientist reflects on her partnership.

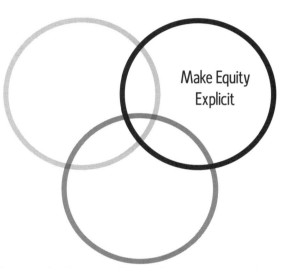

From the retreat that we had at the beginning of the year on, Arthur went into this mega-organizational mode. Which is fine, but it was so funny, from that point on, there were just these roles, you know, these stereotyped roles that were coming out of men and women. Our very first club meeting, my scientist partner, Callie, and I were like, oh, my God, that was the perfect example of how not to do things. It was the perfect example of what we don't want the students to see. Arthur led the first one, I think that was a big mistake. And so there would be comments like, "Oh, Mr. Jenner, will you help me?" The very last thing you want to see. And then in the building activity, we weren't thinking about it, but Arthur was running [the power drill]. Callie was leading the building activity. Arthur was originally supposed to do that, and we realized that wasn't a good idea. So we switched and put Callie on as lead, but then Arthur was running the power drill. It was a two-session activity and after the first class Callie and I both walked out and went, "What were we thinking, you know, it's just stupid. Because both of us can run that station, no problem." And so we told Arthur that, and he was completely fine. He's like, "Yes, of course, of course, of course." I think we were so disappointed that we weren't aware of those things like we should have been. That should have been the first thing we thought of in organizing that activity. It was that, you know, boys should not be running the power drill. Anyone but the boy...

And so I saw all of these things, right, I saw all of these interactions, and I was in the perfect position to say something, but I didn't know how, because some of them were so subtle. And some of them were just so, well, I just didn't know.

Reflection Questions

- What happened with the power drill among this coed team of teacher and scientists? What was troubling about the situation to this female scientist?

SECTION II: Exploring the Triad Framework

- What was the response of the male teacher when confronted about the situation? Was this response surprising to you? Why or why not?
- What might this person have been struggling with in the last line?
- How might this scientist have addressed these issues earlier with her team?
- What are some ways that stereotypes about who can do science are perpetuated in science classrooms? What are strategies for avoiding these stereotypes and addressing them?

Links

- Student Goals: Don't You Feel Powerful?
- Student Goals: The UV Bulb Can Be Changed by the User
- Teaching Goals: My Own Tendency

My Own Tendency

One of the main challenges I faced this year was when I attended our initial retreat for all the teachers and scientists involved in Triad. Females made up the majority of the group. Females were the facilitators. All of the male participants were new to the program since it was the first year the program was going to be coed. The setting was a retreat over a weekend. The purpose was to discuss how to run coed science clubs and clubs just for girls. We had discussions about how to recognize inequitable behavior and what strategies we could use to promote equity in clubs and classrooms.

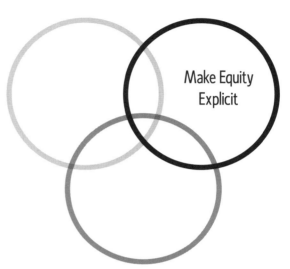

The dilemma for me raised its head when I recognized that our own group was easily falling into the patterns that we were gathered to address. In group discussions, males and a few females (myself included) tended to dominate both in frequency of participation and the length of time spent speaking. It took me a while to recognize what was going on. When I began to notice this pattern I cut back on my own participation and began to observe the group. At this point I was faced with a couple of options. The first option was how to deal with my own tendency to overparticipate. I know that I can easily get carried away and not give others a chance to express their views. My way to deal with that was simple. I reduced my participation. I did not find this frustrating because I was interested in looking at what was happening with the whole group dynamic, so it was easier for me to take the listener/observer role. The second problem was how to bring up this issue with the group and the facilitators or even if I should. I spoke with some other participants, females who also participated actively, and they too had either noticed or quickly became aware of the dynamic our group had fallen into.

I did not talk about what I noticed with our facilitators until the end of the retreat. They seemed to be aware of the dynamic as well, but also not sure how to address it. We briefly discussed some structures/strategies that would have increased overall participation. I noticed that the next time we met as a group the facilitators mentioned that some people tend to participate more than others and that it was important to let all voices be heard.

SECTION II: Exploring the Triad Framework

I found it frustrating and ironic that, as we sat around talking about gender equity, we played out the exact behaviors and roles we were learning to recognize and avoid. With students, I have tried to address equity in a blatant manner. When I notice one group participating more than another—often, but not always, more boys raising their hands than girls, but not always—I stop the group and point out what I have seen. I ask them what they have noticed, if anything, and if they think it matters and why? Then we talk about what we can do to address what the group has identified as an important issue. I felt really weird about the fact that we did not talk about this in our supposedly enlightened group. I think that the matter could have been handled as an "aha" learning moment without people feeling putdown or bad. I did not bring up the matter myself to the group because I did not feel comfortable doing so. This is the most interesting part of the conundrum for me. I often overparticipate. I tend to hang out with others like this. I have a friendly relationship with the males who were over participating. So why didn't I say something? Should I have said something?

Reflection Questions

- Can you share an example of a time when you were aware of an inequity but were unsure if or how you should address it?
- Do you tend to participate very little or a great deal in discussions? How does it feel? What are your reactions to others who participate at different levels than you do during discussions?
- What besides gender could contribute to inequitable working relationships among adults?
- How can we as adults discuss and be explicit with one another about equity issues that arise?

Links

Student Goals: The UV Bulb Can Be Changed by the User

Teaching Goals: No One Felt Uninvolved

Teaching Goals: A Different Role

Teaching Goals: Many People Got a Chance

CHAPTER 5: Teaching Goals

> I made a poster (using cats as students so as not to emphasize gender) that illustrated an inequitable scenario. One cat had all the tools and materials while the other two cats were either sleeping or sad. The poster said, "Practice Equity." We wanted "Practice Equity" to be the Triad mantra. We hung this poster up on the wall so that we could emphasize this message throughout the year by pointing to the poster. Interestingly, I discovered during our initial conversation with the students that they didn't even know what the word equity meant. Now they do.
>
> —*Triad Scientist*

SECTION II: Exploring the Triad Framework

Figure 5F

Teaching Goals and Strategies

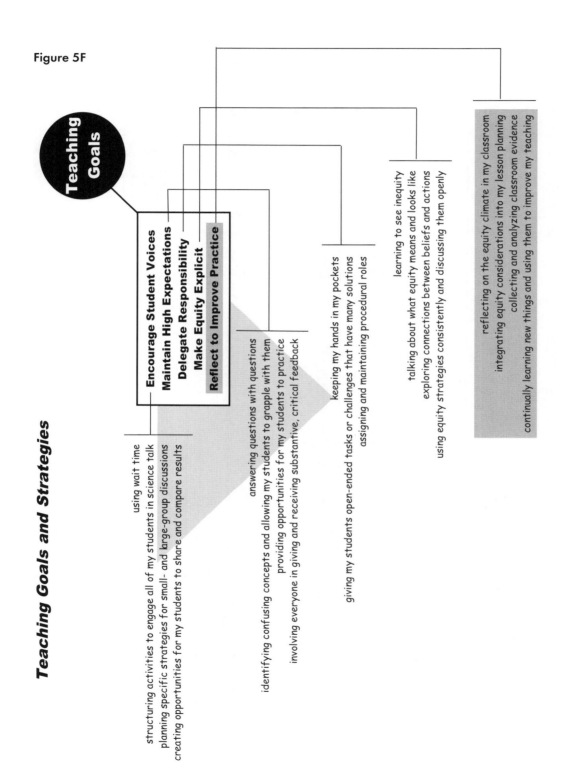

Teaching Goals:
- Encourage Student Voices
- Maintain High Expectations
- Delegate Responsibility
- Make Equity Explicit
- Reflect to Improve Practice

Encourage Student Voices:
- using wait time
- structuring activities to engage all of my students in science talk
- planning specific strategies for small- and large-group discussions
- creating opportunities for my students to share and compare results

Maintain High Expectations:
- answering questions with questions
- identifying confusing concepts and allowing my students to grapple with them
- providing opportunities for my students to practice
- involving everyone in giving and receiving substantive, critical feedback

Delegate Responsibility:
- keeping my hands in my pockets
- giving my students open-ended tasks or challenges that have many solutions
- assigning and maintaining procedural roles

Make Equity Explicit:
- learning to see inequity
- talking about what equity means and looks like
- exploring connections between beliefs and actions
- using equity strategies consistently and discussing them openly

Reflect to Improve Practice:
- reflecting on the equity climate in my classroom
- integrating equity considerations into my lesson planning
- collecting and analyzing classroom evidence
- continually learning new things and using them to improve my teaching

Essay: Reflect to Improve Practice

Scientific knowledge evolves over time through an ongoing process of generating new hypotheses, collecting new evidence, refining our explanations, and working to understand and eliminate contradictions in our theories. In a word, improvement occurs through reflection—reflection on our questions, our methods, our analyses, and our interpretations. It occurs individually, within research groups, across research groups within the same field, and across disciplines. It is, in essence, what the scientific community is all about. It seems logical that improving one's teaching practice might occur through a similar process. The development of new scientific knowledge is by its very nature an iterative process. By *iterative,* we mean that the process is continuous, cyclic, and nonstatic; it feeds back into itself. Improving one's science teaching practice is iterative as well.

We all know that teaching encompasses a wide variety of multifaceted knowledge and skills. Oftentimes referred to as pedagogical content knowledge (Shulman 1986), it includes knowledge of the discipline, instruction, assessment, curriculum, and the specific context in which one teaches. Although books, degree programs, professional-development opportunities, and helpful classroom hints abound, no one can hand you this knowledge in a nice neat package or tell you just how to go about this thing we call teaching once you are in the classroom. You have to be committed to being a learner yourself, and the complex problems and issues that arise on a daily basis are enough to keep the most engaged minds occupied. Dedicating oneself to ongoing inquiry and improvement in one's teaching can lead to confidence based on experience, increased willingness to try new techniques, greater openness to feedback, and increased sophistication in our classroom efforts. An iterative process of professional development is described in detail and strongly advocated in *Designing Professional Development for Teachers of Science and Mathematics* (Loucks-Horsley et al. 2003). As opposed to simply modeling, telling, or scripting what a teacher might do in the classroom, this approach invites teachers to engage themselves and one another in goal-based planning, teaching, collection of classroom evidence, analysis, and reflection. It acknowledges teacher expertise and how it develops, and it brings teachers into an evolving critical process not unlike a scientific research cycle.

Teachers often struggle with finding the support, time, and community resources to engage with others in such work. In Triad we recognized these challenges and tried to build structures that would help us engage with one another as a science teaching community. First, we made time to work together. It will be no surprise to you that this was

SECTION II: Exploring the Triad Framework

our biggest challenge. Then we developed documents we could all use to help us plan, debrief, and share our dilemmas with one another. We articulated goals for our students with respect to both science and equity and then used those goals to guide our planning of classroom investigations. We discussed what kinds of behaviors and student work might constitute evidence that our goals were being met and supported one another in collecting that evidence. We sat down together to analyze our classroom evidence, provided critical feedback for one another, and supported one another in improving our practice. We argued, laughed, cried, and shared our fears and yearnings. We were fortunate to have those times. Although many of us have moved on to other places and positions, we are still working on creating more equitable science classrooms, and many of us continue to talk with one another about our current dilemmas.

Even without a community like Triad, it is possible to engage in reflection on one's teaching practice. We hope the following strategies will be helpful to you.

- Reflect on the equity climate in your classroom. Who likes to get their hands on the materials? Who shares their ideas with confidence? Who enjoys designing their own experiments? Who prefers to have very clear instructions? Who is so passive and compliant that you don't even know much about them? Write down your gut responses to these questions or questions of your own. Then follow up with collecting some evidence.
- Think explicitly about equity as part of your lesson planning. What kinds of groupings will you use? How will you address status issues among your students? What will you do to give students as much time as possible to talk with one another about their understanding and to ensure that each student is developing communication skills? Do you have goals that are specifically related to what you are learning about gender, race, and class and their effects on student opportunities to learn?
- Collect and analyze classroom evidence. If you have access to an audio or video recorder, capture some of the action in your classroom and take time to really see it. Identify some interesting or problematic student work, and share it with a colleague. Perhaps even ask your building administrator for some professional-development release time so you and a colleague can plan a lesson together, and then see what happens when you use it with students.
- Perhaps most important, commit to an ongoing process of learning. Explore both the areas that give you joy and that keep you up at night. Let your students know that you are learning right alongside them. Share some of what you learn with them.

CHAPTER 5: Teaching Goals

You will find more on taking action in Chapter 6. Whether you start with the simple or the complex, your commitment to learning will show in your teaching, and your students will see it. And, hopefully, as they recognize that you are committed to learning, they will also perceive your invitations to explore and understand the natural world as genuine and intended for each and every one of them.

SECTION II: Exploring the Triad Framework

Back in the Classroom

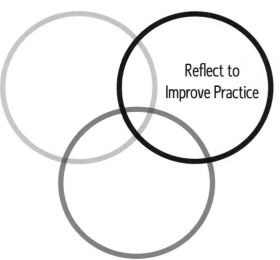

Immediately after science club meetings and team debriefing sessions, I tried to use something from the clubs for my classroom activities. This may have included a new grouping method or a warm-up activity. Because there were one or two Triad club students in my classrooms, they naturally became aides and sometimes leaders in these familiar activities. Back in the classroom, we were now teacher and student, and that meant evaluation and grading. It meant students had to take careful notes and complete all the tasks. But, they didn't have to come up with the one answer, and they didn't have to always have the right answer. They had to show how they came to their answer and be able to reflect. Everyone had something to do. No one was allowed to do nothing. Other familiar new rules for the teacher included: hands in pockets, answer questions with questions, wait for a verbal response, and pick name-sticks (not raised hands) when calling on students.

Reflection Questions

- What specific strategies did this teacher take from her science club experiences back to her classroom?
- What kinds of adjustments do you think this teacher made as she applied what she learned from the science club setting to the classroom?
- What does it take for a teacher to be able to take something new from a learning experience and try it immediately in the classroom? What kinds of learning on your part have inspired you to work in this way?
- From which professional-development experiences have you had the most success in taking the ideas back to your classroom? From which have you had the least success? What do you think made the difference?
- To what extent are you able to debrief with other teachers about new strategies you are trying in your classroom? What do you personally gain from debriefing lessons with others? What, if anything, have you found frustrating about debriefing with colleagues?

CHAPTER 5: Teaching Goals

Links

- Student Goals: The Stopwatch as a Tool
- Science Goals: A Daunting Task
- Science Goals: The Strength of the Group
- Teaching Goals: Answering Student Questions With Questions
- Teaching Goals: I Watched in Awe

By Scoring When a Girl Participated

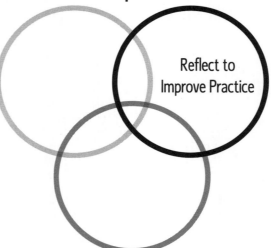

It was a great learning experience to determine what types of group discussions encourage student voices and a high level of participation. On our team, we found that small-group discussions tended to promote the highest level of overall participation and seemed to encourage the girls who tended to be more quiet to speak up and voice their ideas. We implemented these group discussions in a couple of ways—by posing questions to the girls and then asking for answers by raising their hands or simply going around the table and asking for answers from each girl in turn. Through scoring when a girl participated in the conversation, we found that the level of participation was very good in either format. The quieter girls spoke up less in the voluntary setting as opposed to the round-robin discussion; however, they did have a much higher level of involvement than when we discussed science with class as a whole. This helped me to understand the idea of equitable teaching and how it doesn't necessarily fall into a category of race or gender.

Reflection Questions

- The author of this reflection refers to "scoring when a girl participated." Have you scored relative rates of participation among boys and girls in your classroom? If so, what have you found?
- Although she may not know it, this author is engaging in *action research,* that is, she is asking questions about her own teaching practice and collecting classroom evidence to answer her questions. What opportunities, if any, have you had to conduct action research in your classroom? What strategies could you use to collect this kind of data in your classroom while you are also engaged in teaching?
- What observations have you made in your classroom that have caused you to significantly change the way you teach?
- What do you think the author means by the statement, "This helped me to understand the idea of equitable teaching and how it doesn't necessarily fall into a category of race or gender"?

CHAPTER 5: Teaching Goals

Links

◎ Student Goals: I Assumed That Our Girls Would Feel Comfortable

◉ Science Goals: Above a Whisper

◉ Teaching Goals: I Watched in Awe

> I've always wondered how important what I say as a teacher is in promoting understanding versus how important it is for students to hear one another's ideas. I suspect that student voices are more important than I used to think.
>
> —*Triad Scientist*

SECTION II: Exploring the Triad Framework

Personal Development

Creativity was one of our goals for our students. During our team debrief after a science activity about pendulums, it came out that one team member thought that goal meant scientifically creative—as in where and how to attach the pendulum or how to add washers when changing weight—while another team member thought that the goal meant more abstract creativity, such as using pendulums as a model for designing a swing and asked then what sort of things would make the swing fun. As a team, we never figured out how to communicate with one another, so we all had different definitions of the same goals.

Although this may not sound like a positive experience, I saw it as enlightening. I will be working in groups of one kind or another throughout my life, especially since I am in science. I often experience frustration in groups and now see that it often stems from miscommunications like the one described above. With the new knowledge of what might be going wrong in my communication with others, I can work to be more precise and perhaps avoid these frustrations from the beginning. This new understanding will also be valuable to me when I am responsible for overseeing a group—for instance, when I run a laboratory of my own. Having worked through frustrations associated with group work myself and learned techniques for working with others efficiently, I can help my lab members find better ways to work together.

I joined Triad to help middle school girls develop a love of science, but what I really got out of Triad was personal development. I learned new ways to interact with people, observed how these philosophies and guidelines could lead to an efficient, happy, organized group working toward a common goal; and I became aware of my own strengths and weaknesses associated with group work and teaching.

CHAPTER 5: Teaching Goals

Reflection Questions

- What did this scientist learn personally from her reflections on her communication with her team?
- Through reflection, this individual is taking lessons learned from working in a scientist-teacher partnership and applying them professionally to her work as a scientist. Has reflection on your own teaching experience ever led to a personal development on your part? Describe the circumstances and what you learned.
- This teaching team struggled because it did not share a common understanding of its goals for students. Do you think it's important for your students to understand your goals for them? As a community of scientists, how might you and your students reach a common understanding of your goals?
- This scientist is reflecting on her experiences working in a group and predicting the impact it will have on her professionally in the future. What kinds of personal learning experiences have you had working in groups? How has this influenced your teaching? How have you helped your students learn to navigate difficult group dynamics?

Links

- Student Goals: So That All the Bridges Fall
- Student Goals: On a More Personal Level
- Science Goals: Not What We Had Planned
- Teaching Goals: A Different Role

One teacher found that personal development was the most important thing she took away from Triad.

SECTION II: Exploring the Triad Framework

Resurrecting Socrates

Utilizing the reflection process this year was of tremendous assistance, as it is whenever I make time to incorporate it into my classroom teaching. I'm still learning when and how to use reflection effectively, though I always have insights when I make the time. For example, in our culture of memorization as learning and of standardized test scores as a measure of student success, it can be difficult to resurrect Socrates. Learning to ask questions of value, questions that teach students what and how they're thinking, is such an empowering experience for both the student and the teacher. I realized upon reflection that I can easily lose sight of questions in the rush to cover all the material. Triad gives me the opportunity to practice questioning: purposeful questioning, thoughtful questioning, deliberate questioning. It seems somewhat incongruous to me that, of all the disciplines, it would be such a challenge to learn how to ask questions in science. Although some of that challenge really seems, at its heart, to be about trusting students to question and trusting them to join the teacher in exploring questions, rather than expecting canned answers.

Reflection Questions

- How do you make time to incorporate reflection into your teaching practice? When do you reflect? What does it look like?
- Reflection often results in new ideas for teaching. Given the pressure to "cover all the material," how do we try out these new ideas and make changes in the classroom?
- In what ways is the practice of questioning an essential part of reflection?
- How do you think students' desires for canned answers are related to their experience in science classrooms? Reflecting on your own use of questions in the classroom, to what extent do you think the kinds of questions you use contribute either to your students' comfort in asking and exploring versus needing simple answers?

CHAPTER 5: Teaching Goals

- Reflect on your own use of questions in the classroom. For what purposes do you ask questions? What kinds of questions do you ask—questions with multiple possible answers or one answer? How do you ask most of your questions—in writing, verbally to the whole class, or verbally with individuals? To whom do you address questions—the whole class, small groups of students, individual students, yourself?

Links

- Student Goals: So That All the Bridges Fall
- Science Goals: Answers Are Not the Goals
- Science Goals: To Simply Marvel
- Teaching Goals: Does This Bridge Look Better Than It Did the Last Time?

> You don't want to cover a subject; you want to uncover it.
> —David Hawkins

At First I Was Hesitant

At first I was hesitant to do the mobile activity, because I felt that it was sneaky doing an equity activity in a science club. After seeing the results, however, I now realize that the students were game. Most groups realized the inequitable distribution of materials about halfway through the activity. This was mainly due to the disruptive behavior of one of the groups that had very few resources and went from group to group to complain. They finished very quickly, because they didn't care, and they were mad. They named their mobile the "Garbage Mobile." On the other hand, the other group with very few resources treated the inequity as a challenge. They were going to make the best darn mobile with what they were given. It turned out to be very clever, and they named it the "Eat Your Vegetables" mobile. There were two different outcomes with the groups who were given the most resources, as well. One group made a very colorful, smart mobile made of different origami animals, while the other group, overwhelmed by the amount of supplies, couldn't finish in the allowed time. This was the only group that was unable to finish a mobile.

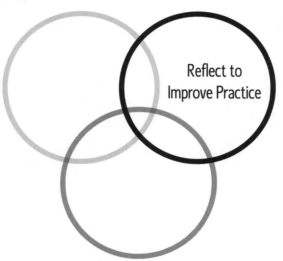

Afterward, we discussed equity with the students. Questions arose, for example, about the fairness of judging these mobiles when it was clear that not everyone had access to the same materials. We discussed affirmative action, getting into the best high school, and the annual science fair competition. In other words, we discussed many situations in which people are judged on equal grounds regardless of access, whether it be access to an available knowledge base, access to materials, or access to money, time, or advice. If I were to do it again, I would bring a video camera (maybe more than one), and I would have the discussion in a more intimate setting, perhaps in small groups.

Reflection Questions

- Describe a time you took a chance and taught something you were initially hesitant to try. What happened? Was it worth it? Would you do it again?
- Students come to the classroom with very different access to a variety of resources. How might reflecting on the differing responses described above among the low-resource and high-resource groups in the mobiles activity help you, as a teacher, to work effectively with students who respond differently to their own access to resources?
- This teaching team involved their students in reflecting on the mobiles activity and its parallels to everyday life. How can student reflection provide information to teachers that can help them improve their practice? Describe a time that a student's reflection on his or her learning changed the way you teach.
- The author states, "If I were to do it again, I would bring a video camera." Have you had the opportunity to videotape your classroom in action and watch it later? If so, what did you learn from this experience? If not, how might you be able to make such a recording so that you could examine your classroom in this new way?
- After teaching a lesson, how often do you think to yourself, "If I did it again, I would …" How do you tuck these insights away so you can revisit them in the future?

Links

- Student Goals: After the Initial Eeewwww
- Science Goals: I Learned How a Lava Lamp Works
- Teaching Goals: Way Beyond Our Expectations
- Teaching Goals: Stop in My Tracks
- Teaching Goals: Accepting Stereotypes

SECTION III:
Looking Forward and Learning More

CHAPTER 6

Taking Action

Now that you have pondered the ideas presented in this book, how might you take action yourself? In the spirit of Triad, we hope this book will spark ideas for what steps you could take in your own context. To provide some guidance, we have outlined in this closing chapter five possibilities for how to get started in taking action yourself to promote gender-equitable science teaching: Make the Triad Framework Your Own, Reflect Through Writing, Engage in Classroom Research, Start an After-School Science Club, and Build a Reflective Community.

Make the Triad Framework Your Own

The Triad Framework presented in this book was developed from our experiences over several years working with teachers, scientists, and middle and upper elementary school students in after-school science clubs. The Triad Framework began with small steps and grew slowly. Triad teachers and scientists used the Framework as they planned their science-club lessons together, and teachers applied the Framework to their classrooms as well. With feedback from teachers and scientists and our own insights from working with the Triad community, the Framework evolved and changed along the way. Over the years, goals were rephrased, reordered, added, and removed. The suggested strategies for each goal also went through several iterations. Teachers and scientists, when planning their lessons together, would often develop their own additional Student, Science, and Teaching Goals.

We invite you to continue this evolution—to make the Triad Framework your own and to apply it to your specific context. The Triad Framework was undoubtedly influenced by the setting in which we developed it: the diverse elementary and middle schools of the San Francisco Unified School District (SFUSD) and the biomedical sciences community of the University of California at San Francisco (UCSF). That said, we have since used the Framework in a variety of settings at new institutions with new colleagues and contexts, and we invite you to do the same. Although the Triad Framework is designed as a tool for gender-equitable science teaching, it can be adapted and used to address other equity issues in the classroom and other settings. For example, how is the Framework relevant and useful when working with special education students? English language learners? African American students? Native

SECTION III: Looking Forward and Learning More

American students? High school or even college students? How would you change the goals of the Framework when working with these groups? What would work best with these students? What, if anything, would you need to add from the Framework? What, if anything, would you remove?

For example, the Triad Framework has been adapted for use with elementary school teachers and their English language learner students. In this context, the Teaching Goal Encourage Student Voices takes on a special significance. The strategies listed—using wait time, structuring activities to engage all students in science talk, planning specific strategies for small- and large-group discussions, creating opportunities for students to share and compare results—are all relevant and useful for English language learners. That said, many English language learners need additional support before they can voice their thoughts and/or results. One can thus imagine adding strategies or making some of the listed strategies more specific. What are ways of "structuring activities to engage all of my students in science talk" when working with non-native speakers? One possibility is to provide your students with sentence frames; examples are: "The cow's eye felt _____" and "When we bounced the different balls, I noticed that _____." The amount of detail provided in the sentence frame depends on students' English language fluency. English language learners will benefit from vocabulary support—they may not be able to share their results without such support. There are many ways to do this, including word walls (posting words relevant to the lesson(s) in the classroom), word list handouts (for example, a list of relevant adjectives for "The cow's eye felt _____."), and labeling materials and diagrams.

As we suggested in Chapter 2, we hope that you will sit down with the Framework and whatever tools you desire—pens, Wite-Out, wine, watercolors—and edit it to make it more useful and relevant for you and your students. Depending on the context in which you teach, these edits might be small or large. For example, we could imagine adapting the Framework to other subjects, such as language arts, mathematics, and social sciences. Such an adaptation would likely involve significant changes. Finally, we hope that you will return to the Framework again and again. As we have done so, we have found that our understanding and use of it has changed—and we have continued to edit, using both real and metaphorical Wite-Out and word processors, to change goals and strategies so that they align with our current and ever evolving thoughts on the dilemmas of equitable science teaching.

CHAPTER 6: Taking Action

Reflect Through Writing

In the fast-paced world in which we live, we rarely take or have the time to reflect through writing. Yet doing so can be incredibly valuable, as evidenced by the dozens of provocative dilemmas that emerged from the reflective writings of Triad teachers and scientists presented as vignettes in this book. We found that these reflections helped us learn—about partnership, teaching, learning, science, students, equity, and more—and we hear from our participants that the reflections helped them to learn as well—about themselves. In the words of one teacher:

I also learned the importance of giving students the time to reflect. This idea was mentioned in the article on science journals but ultimately hit home during my own reflections. It was during these reflections that I made connections about presented material that had not occurred to me at the time it was presented. The reflections also led to questions that had not previously occurred to me.

The majority of the vignettes shared in this book were pulled from reflections written by Triad teachers and scientists in response to the prompt, "What did you learn from your Triad experiences both in and out of the club setting? How did you learn these things?" If you would like to do some writing but are unsure where to start, consider using a variation of this prompt, for example, "What did you learn from your experiences today, this week, this month, this semester, or this year as a teacher? As a professor? As a student teacher? As a scientist volunteer? As a professional developer? How did you learn these things?" Give it what time you can. With more time, you can expect to delve more deeply into your successes and struggles, but even small amounts of time—as little as 30 minutes—likely will draw out salient dilemmas. Ideally, you might find a time or a place that will be free of interruptions from such things as e-mail and phone calls.

Material ripe for us to ponder is ever present. Some of us take the time to reflect at the beginning or end of each day. Turning points, such as the beginning or end of a school year or the first time we teach a course or classroom unit, also offer reflection opportunities. A vignette in this book may have raised questions for you that you would like to explore through writing. Sometimes we lay awake at night thinking about our work—in essence, a dilemma is demanding time from us. Perhaps taking the time to reflect through writing will help. However it happens, whatever its inspiration, we encourage you to reflect through writing.

SECTION III: Looking Forward and Learning More

Engage in Classroom Research

When you think about your own classroom and students, what questions do you have? Perhaps you would like to know more about verbal participation—the rate at which different students share verbally, how participation rates change in whole-group and small-group settings, and whether doing a think-pair-share with your class increases the participation or influences who speaks are just a few of the many questions one could investigate about students' verbal participation. Maybe you are interested in finding out which students have their hands on the materials the most. The questions we can ask about our teaching and our students are endless.

> **Think-Pair-Share**
>
> Think-pair-share is a teaching strategy that encourages the engagement and participation of every student. Its steps are:
>
> **Think:** The teacher poses a question. Each student has time to think about the question—in her or his head or perhaps by jotting down some notes.
>
> **Pair:** Students discuss their ideas with each other: This sharing is usually done in pairs so that each student is likely to have the opportunity to talk about his or her ideas. This sharing can be unstructured or each student can have a timed amount of time to talk while the other student is expected to listen.
>
> **Share:** Students now share their ideas with the rest of the class—the structure for this sharing could be that each pair shares, people raise their hands to share, or some other system of your choice.

As a place to start, think about your own interests, perhaps those that have emerged through written reflection, and brainstorm how you could gain some insight into them. How systematic your investigation is is up to you. Sticking with the verbal participation example, perhaps you want to start by simply keeping track of which students speak in the whole-group setting on a given day. Such an investigation would be relatively straightforward. And perhaps doing so would pique your interest, making you want to learn more. A likely next step is to collect some more data—perhaps by doing the same procedure on a different day or keeping track during small-group discussions. Another possible next step is to explore the literature related to the topic—what others have said about it and how have they investigated it. (*Lesson Study: A Handbook of Teacher-Led Change* by Catherine Lewis [2002] is one good place to start.) Both of these steps will help you to decide how you want to move forward.

Although how you move forward will be your decision, many educators suggest that you

- select a question,
- collect, organize, and interpret information related to the question,
- study the relevant professional literature and research,
- determine possible actions, and
- take action and document results (Calhoun 1994).

In choosing a question, pick one that interests you and one whose results could have an impact on teaching and learning practices in your classroom. Your first data collection will probably be mostly exploratory—gathering information from your classroom that will inform your research design or next steps. Reviewing the relevant literature can likely be done mostly online; your town or city library and local university libraries may also be helpful. Now that you have gathered more information, you can decide how you want to move forward—what kinds of investigations you will do in your classroom and how you will keep track of your results. Possible sources of data are observations of your classroom, interviews with students, and student written work. Often times, using more than one source of data and repeated observations will provide you with greater confidence in your results. After you gather your data, take some time to review it and decide on your next steps. Triad participants found that a particularly useful manual for them was *Teachers Investigate Their Work: An Introduction to Action Research Across the Professions* (Altrichter et al. 1993).

Research in your classrooms can be aided by your colleagues—perhaps you can ask a colleague you trust to make observations in your classroom to help you answer your question(s). Ideally, you would share your question with your colleague and your expectations for data collection. That is, what kinds of things do you want him or her to look for? What would you consider evidence related to your question? If another teacher is unavailable, Triad participants have also used audio or videotape to capture the classroom interactions that are fleeting and difficult to observe while in the midst of teaching. See Karen Gallas's *"Sometimes I Can Be Anything": Power, Gender and Identity in a Primary Classroom* (1997) and *Talking Their Way Into Science* (1995) as examples of how one teacher did research in her classroom using field notes and audiotapes.

SECTION III: Looking Forward and Learning More

Start an After-School Science Club

Another possibility is to work with a colleague or two to start an after-school science club that can serve as an equitable-teaching laboratory in which you can try new strategies and assess how they worked, as well as provide after-school enrichment for students. Working in the after-school environment frees you from many of the constraints of the classroom, and, by partnering with at least one other person you have more opportunities to observe the "classroom" dynamics. Ideally you will partner with at least one other person, but you could also consider leading a club by yourself.

A suggested place to start as you plan your club meetings is the Triad Framework. Determine which Student, Science, and Teaching Goals you will focus on (adapted as desired for your context) and then design a science club that will help you meet them, taking into account your knowledge of your students. You'll likely want to think about student groupings, the structure of the activities and how students will be both challenged and supported, roles and responsibilities during the club meeting, the introduction to and wrap-up of the club, and so forth. You'll also want to think about what will give you evidence that you are meeting or working toward your Student, Teaching, and Science Goals.

In addition to exploring the Framework's goals and strategies and their impact, the club provides a good opportunity for posing your own question(s) and exploring them in an after-school environment with more flexibility, maybe fewer students, and ideally at least one other colleague. Do you have other strategies that you want to try out? What questions are you interested in exploring? How will you and your colleague(s) go about gathering evidence to answer your question(s)?

After a club meeting, be sure to include time to reflect with your colleague(s). If you are leading the club solo, we encourage you to spend some time writing notes and reflections after a meeting. Consider the extent to which the activity design and the strategies addressed their stated goals, and how effectively they were implemented. Out of this reflection new questions and ideas will emerge that can inform your next steps: What could be changed next time to more effectively implement these strategies? Begin planning your next club session taking into account what you have now learned.

You can sponsor this after-school "equity laboratory" with any interested colleague, but you might also be interested in finding a local scientist or two to join you. Possible ways for connecting with local scientists are (1) finding a partnership program hosted by a local college or university through searching science department websites

CHAPTER 6: Taking Action

or calling the science departments; (2) asking your colleagues if they know of scientist parents or other scientists who might be interested in jointly leading an after-school science club; and (3) contacting scientific societies to see if they have members in your area who are interested in volunteering in science education. Coplanning and coleading science clubs can be greatly rewarding for all involved: the students, the scientist, and the teacher. However, if you're partnering with someone you've newly met, you'll likely want to spend extra time getting to know one other and discussing club goals, roles, responsibilities, and how you will gather evidence regarding your questions.

The structure of your clubs will be for you and your colleagues to determine. That said, perhaps the Triad model can provide you with a place to start. In Triad, one to two teachers partnered with two scientists (mostly graduate students in the biomedical sciences) to lead a series of after-school clubs. They also attended a yearlong professional-development series that focused on gender-equitable science teaching and reflecting on our own and others' practice. The clubs met about twice a month from about December through May for a total of 10 times, and each meeting lasted for one-and-a-half to two hours. Each club session was designed by the club sponsors—the teachers and scientists—and the science activities were varied and included a range of science topics.

Build a Reflective Community

This book was a collaborative effort that grew out of a community of teachers and scientists who worked together for many years. The Triad teachers and scientists grew more reflective over the years, supported in these efforts by the Triad staff and the evaluation team. A reflective, collaborative, and supportive community is an ideal place to use this book, explore its ideas, develop your own—individual and group—ideas, and pursue the previously described suggestions as desired. Such communities might be found within a class (preservice or in-service), a professional-development program, or a school's faculty.

Vignette discussions are a natural place to start in building a reflective community. The heart of this book consists of such vignettes—examples, reflections, and stories that illustrate a person's underlying beliefs. Our beliefs about our actions are central to our teaching and to making changes in our teaching. By discussing a vignette with others, we have the opportunity to articulate our beliefs and actions, gain insight from other perspectives, learn from others, and share and receive feedback. Un-

SECTION III: Looking Forward and Learning More

doubtedly one will uncover new meanings in a vignette when talking about it with others, even if it is a vignette one has talked about before. Vignette discussions enable participants to create and develop their own philosophies of practice while considering the philosophies of others. They can also provide opportunities to bridge theory and research with actual practice. Vignettes privilege the experience and knowledge of the practitioner rather than the researcher. If you will be facilitating a discussion of a vignette, we encourage you to read our suggested guidelines in the appendix.

The Triad Framework can also be used as a tool to think about access and equity issues and grapple with these ideas with others. We have found that, for many, the Triad Framework provides an accessible entry point for thinking about gender-equitable science teaching. Consider sharing the Framework with some colleagues as a way to dive into a discussion of these and related issues. You can talk about the goals and their strategies—for example, sharing times when you've used such strategies and the resulting outcomes, other strategies you would add to a particular goal, and the overlap and interplay of the Student Goals, Science Goals, and Teaching Goals.

Conclusion

We hope these suggestions provide ways for you to move forward and take action toward equitable science teaching. Such work is ongoing and iterative and with each step we take we learn, grow, and gain new insights into how to make science more accessible to all students.

APPENDIX A

Facilitation Guidelines

This appendix provides suggestions to keep in mind when discussing the vignettes presented in this book with others and, in particular, when facilitating a group discussion using a vignette. Places where such discussions could occur include a professional-development workshop, among a group of teachers at a school site or within a district, in a college course, or at a conference.

Before the Discussion

Each vignette is accompanied by a set of questions—these questions were designed to be thought provoking and to stimulate group discussion. The questions are a great place to start when talking about a vignette with others—we, of course, encourage you to think of your own questions—but, before you get to the questions, it's also important to think about your participants, your goals for the discussion, and your role as facilitator.

First, consider your participants. Is this a group that knows one another well? Is it the group's first time together? The group's only time together? What are the teaching backgrounds of those in the group? What are the science backgrounds? How similar do you expect participants' experiences and opinions to be? A new teacher may not have as much practical in-class experience from which to draw as an experienced classroom teacher. On the other hand, a new teacher fresh from a credentialing program may be more familiar with recent educational research. A new high school science teacher may not feel confident commenting on classroom or teaching issues but may be very confident when discussing science content or process. The unique background of individuals will influence their participation in the discussion—shaping their perspective, influencing their likelihood of sharing, and affecting how their views are received and perceived by others. In this respect, a facilitator might also consider each person's profession, education, native language and skill with the English language, race/ethnicity/culture, gender, age, and other characteristics.

As you think about your participants, also reflect on your goals for the discussion. Are your goals for the discussion more at the individual or group level? For example, at a conference, you would probably be more focused on helping individuals use the

APPENDIX A

discussion to think about their own beliefs and practices, while, in a course or professional learning community, you would likely want participants also to be learning about each other and building a community. Additional goals might include relating participants' previous experiences to current contexts, using the discussion as a springboard for setting some common ground or planning a group project, and having a philosophical discussion that delves deeply into potentially difficult subjects, such as stereotyping.

A vignette's focus will likely influence who participates and what that participation looks like. Think of participation in a broad sense—including both talking and listening. As an example: If a vignette uses a student's voice, every participant might be more likely to comment, because we have all been students. On the other hand, if the vignette is written from a scientist's perspective, a teacher might feel less comfortable in making a comment. If a participant strongly identifies with the particular situation represented in a vignette, he or she might feel particularly vulnerable while others discuss and critique the vignette. A skillful facilitator can anticipate such vulnerability and create a safe place for participants to be explicit about their beliefs and actions by being familiar with participants, setting a tone for respectful disagreement, and emphasizing norms (see "Facilitating the Discussion," which follows) that ensure all participants have opportunities to contribute and learn. Finally, if participants do not know each other well, they may be less willing to disagree. To challenge one another may seem a bit abrupt. A facilitator can help by first providing clear expectations for the discussion and then by asking questions designed to break consensus.

Facilitating the Discussion

The learning that occurs during a vignette discussion is in part dependent on how the discussion is facilitated. When we have used a vignette with groups, we have found it very useful to establish group norms and facilitate the discussion. When we facilitate discussions, we find it helpful to decide ahead of time which equity strategies we will use, introduce ourselves to the group before the discussion begins (sounds obvious but can be easy to forget), and keep time or ask someone to keep time for the group. We then share group norms at the beginning of each discussion. These norms have included that each person will contribute to the discussion and listen to the contributions of others; in sum, each person will stay engaged. We also expect everyone to be open and honest when sharing ideas, to be respectful of the variety of perspectives that are shared, and to maintain confidentiality if applicable. Most likely, if these norms are followed, each person will experience discomfort at some point during the discussion. To help the discussion move smoothly, we ask that each person help maintain a focus on the discussion topic and that they direct their questions and comments to the group members and not the facilitator.

After discussing the group norms, we review our role as facilitators with the group. For example, one of us might say, "As the facilitator, I will pay attention to the group norms we've just discussed and help the group maintain them. I will also remind people, as necessary, to speak to the group and not to me. I will try to refocus the group if we stray from the discussion topic and will prompt the group to go on to the next topic or question when appropriate. I will ask for clarification as needed and will keep track of time. Finally, it is not part of my role as facilitator to comment on everything that is said."

Several possible statements might be helpful during the discussion. One prompt for refocusing the group is "The issues we're talking about are important, but we've gotten away from the discussion topic/question. Let's return to _____." To suggest the group move on, one might say, "The issues we're talking about are important, but we have only_____ minutes left and we still need to talk about _____." or "We have spent a lot of time on this particular point, and, while the conversation is valuable, we need to move on to _____." To promote verbal participation by all group members, near the midpoint, if not everyone has spoken, make that observation by stating, "We have _____ minutes remaining and we haven't heard from _____ or _____," or "Our conversation will be richer if everyone shares their ideas. We haven't heard from _____ or _____ yet."

In any situation, the facilitator may need to consider possible methods to break consensus among group members, such as offering alternative points of view or asking provocative questions. Group members may overwhelmingly agree or not challenge one another on sensitive subjects—they may be somewhat preoccupied with getting to know one another, focused on gaining an understanding of the process, or aware of status issues within the group. Perhaps the vignette is not particularly provocative or relevant to them upon first glance. Asking questions such as "Do you agree with this? Disagree? Why? What would you do in this situation? Does anyone have a different opinion? Can anyone imagine a different perspective?" and using wait time (Rowe 1974) will help participants feel more comfortable sharing different viewpoints. By presenting alternative opinions, the facilitator may help to spark more exploratory conversations. Often, if the facilitator presents more challenging questions to be addressed, participants may feel more at ease commenting on them than if a fellow group member had presented them.

APPENDIX A

Finally, to what extent is facilitation akin to teaching? In other words, facilitation is in part a matter of personal style and each of us needs to find a facilitation style with which we are comfortable. As in any situation when humans are placed together and asked to deliberate about difficult issues, there are many variables to consider. No one particular approach will guarantee any one particular outcome. We have simply tried to share strategies that we have found useful in many different settings with various groups over the years.

APPENDIX B
A Few Words on Data Collection and Methodology

Triad Staff Methodology

Increasingly, over the course of the history of the project, the Triad staff viewed evidence collection and program evaluation as an integral part of science education partnership work and professional development rather than as a segregated endeavor or external process of review. Evaluation data was collected and analyzed regularly as part of professional-development activities and staff planning efforts, resulting in collaborative analysis, reflection, and strategizing for action. The data presented in this book is derived primarily from evaluation evidence collected from the Triad community of teachers and scientists. The majority of the evaluation data presented in the core chapters of the book are excerpts from culminating, end-of-the-year reflections of Triad teachers and scientists written in response to the following open-ended prompt: "What have you learned this year from your Triad experiences in and out of the classroom setting. How have you learned these things? Please provide supporting evidence for your statements." This probe, while seemingly simple, proved reliable in eliciting rich descriptions of participants' struggles and achievements. It captured the breadth and depth of participants' ideas on their own learning and did so in their own voices. In addition, teacher and scientist participants were supported and encouraged in their efforts to collect and analyze data from the students with whom they worked, including student responses to prompts such as "Draw a picture of your favorite Triad activity," and "Why was this your favorite activity?" We selected the participant data in the core chapters of this book not to convey the successes or outcomes of Triad but rather to highlight the questions, dilemmas, and challenges that arose so that others can discuss them during their own learning about gender equity in science education. We took care to remove any identifiers from these excerpts and gain permission for use of these reflections in the book.

Stanford Evaluation Team Methodology

The role of the Stanford Evaluation Team was originally conceived as a critical friend that could see the project through a different lens and assist the Triad Staff in continually and iteratively improving the program. (The evaluation was conducted under terms of a subcontract from the University of California at San Francisco [UCSF 2108SC]).

APPENDIX B

The Stanford Evaluation Team agreed to observe Triad activities, interview adult and student participants, discuss program issues with the staff, and generally highlight issues and challenges for the Triad Staff that, if addressed, would enable the project to better achieve its goals. This evaluation effort was not designed to pronounce judgments about the quality of Triad's work, though considerations of quality inevitably permeated virtually every discussion between Triad staff and the evaluation team. It quickly became apparent, however, that the Triad staff itself was already engaged in systematic and careful analysis of its own program—and making changes as a result. From the outset then, the Stanford Evaluation Team saw itself more as a partner in a self-study than as a group of dispassionate outsiders issuing authoritative opinions.

Data collection by the Stanford Evaluation Team involved a mixed-method approach, incorporating elements of quantitative and qualitative traditions and including surveys, field observations, interviews, focus groups, and audio and video recordings. Observations were made of all components of the Triad project: scientist orientation workshops; after-school club meetings with students; planning meetings involving teachers, Triad staff, and scientists; professional-development workshops; staff planning meetings for professional-development workshops; debriefing meetings; and retreats in which all adults participated. The observations were intended to highlight emerging issues within Triad and to help investigate links between the themes and goals of the project and its actual operation. Interviews of individual teachers, scientists, and students were conducted regularly throughout the project. Debriefing sessions and written reflections provided valuable formative feedback throughout the project. Data were collected following each professional-development workshop and community retreat, as well as at year's end in the form of final reflections written by all adult participants.

As the involvement of the Stanford Evaluation Team came to a close, attention shifted toward how best to share the wealth of evidence and insight that had been collected over the years. It was decided that the core activity involving the Stanford team during the final year would be the development of a sourcebook that might help teachers identify and examine issues of gender equity in their own classroom settings. The preparation of such a sourcebook exemplified the close collaboration that had been solidified between the Triad staff and the Stanford Evaluation Team, and this book is the culmination of that collaboration.

APPENDIX C

Literature Cited

Adler, R. 2007. The curse of being different. *The New Scientist* 2586: 17.

Altrichter, H., P. Posch, and B. Somekh. 1993. *Teachers investigate their work: An introduction to action research across the profession.* London: Routledge.

American Association for the Advancement of Science (AAAS). 1989. *Project 2061: Science for all Americans.* New York: Oxford University Press.

American Association for the Advancement of Science (AAAS). 1993. *Project 2061: Benchmarks for scientific literacy.* New York: Oxford University Press.

American Association for the Advancement of Science (AAAS). 2001. *Project 2061: Atlas of scientific literacy, Volume I.* New York: Oxford University Press.

American Association for the Advancement of Science (AAAS). 2007. *Project 2061: Atlas of scientific literacy, Volume II.* New York: Oxford University Press.

American Association of University Women (AAUW). 1992. *How schools shortchange girls.* New York: Marlowe.

American Association of University Women (AAUW). 1998. *Gender gaps: Where our schools fail our children.* Washington, DC: American Association of University Women Educational Foundation.

American Association of University Women (AAUW). 2000. *Educating girls in the new computer age.* Washington, DC: American Association of University Women Educational Foundation.

Aronson, J., C. B. Fried, and C. Good. 2002. Reducing the effects of stereotype threat on African American college students by shaping theories of intelligence. *Journal of Experimental Social Psychology* 38 (2): 113–125.

Black, P., and D. Wiliam. 1998. Inside the black box: Raising standards through classroom assessment. *Phi Delta Kappan* 80 (2):139–148.

APPENDIX C

Black, P., C. Harrison, C. Lee, B. Marshall, and D. Wiliam. 2004. Working inside the black box: Assessment for learning in the classroom. *Phi Delta Kappan* 86 (1): 8–21.

Bolgatz, J. 2005. *Talking race in the classroom.* New York: Teachers College Press.

Brickhouse, N., P. Lowery, K. Schultz. 2000. What kind of a girl does science? The construction of school science identities. *Journal of Research in Science Teaching* 37 (5): 441–458.

Bronson, P. 2007. How not to talk to your kids: The inverse power of praise. *New York Magazine,* from http://nymag.com/news/features/27840.

Calabrese Barton, A., with J. Ermer, T. Burkett, and M. Osborne. 2003. *Teaching science for social justice.* New York, NY: Teachers College Press.

Calhoun, E.F. 1994. *How to use action research in the self-renewing school.* Alexandria, VA: Association for Supervision and Curriculum Development.

Catsambis, S. 2005. The gender gap in mathematics: Merely a step function? In *Gender differences in mathematics: An integrative psychological approach,* eds. A. M. Gallagher, and J. C. Kaufman, 220–245. New York: Cambridge University Press.

Chubb, N. H., C. I. Fertman, and J. L. Ross. 1997. Adolescent self-esteem and locus of control: A longitudinal study of gender and age differences. *Adolescence* 32 (125): 113–29.

Cohen, E. G. 1994. *Designing groupwork: Strategies for the heterogeneous classroom.* New York: Teachers College Press.

Cohen, E. G., and R. A. Lotan, eds. 1997. *Working for equity in heterogenous classrooms: Sociological theory in practice.* New York: Teachers College Press.

Driver, R., A. Squires, P. Rushworth, and V. Wood-Robinson. 1994. *Making sense of secondary science: Research into children's ideas.* New York: Routledge.

Duckworth, E. 1996. *The having of wonderful ideas and other essays on teaching and learning.* 2nd ed. New York: Teachers College Press.

Dweck, C. S. 2000. *Self theories.* Philadelphia: Taylor and Francis.

Dweck, C. S., and E. L. Leggett. 1988. A social-cognitive approach to motivation and personality. *Psychological Review* 95: 256–273.

Ehrlich, P. R., D. B. Dobkin, and D. Wheye. 1988. *The birder's handbook: A field guide to the natural history of North American birds.* New York: Simon and Schuster.

Fenema, E. 2000. *Gender and mathematics: What is known and what do I wish was known?* Paper presented at the Fifth Annual Forum of the National Institute for Science Education.

Finson, K. D. 2002. Drawing a scientist: What we do and do not know after fifty years of drawings. *School Science and Mathematics* 102 (7) 335–345.

Gallas, K. 1995. *Talking their way into science: Hearing children's questions and theories, responding with curricula.* New York: Teachers College Press.

Gallas, K. 1997. *"Sometimes I can be anything": Power, gender, and identity in a primary classroom.* New York: Teachers College Press.

Gardner, H. 1983. *Frames of MIND: The theory of multiple intelligences.* New York: Basic Books.

Gatta, M., and M. Trigg. 2001. *Bridging the gap: Gender equity in science, engineering, and technology.* New Brunswick, NJ: Center for Women and Work.

Gordon, J., ed. 1995. *I did it! An educator's guide to developing mastery-oriented learners.* Denver, CO: Girls Count.

Holden, C. 1993. Giving girls a chance: Patterns of talk in co-operative group work. *Gender and Education* 5: 197–189.

Kivel, P. 2002. It's good to talk about racism. In *Uprooting racism: How white people can work for racial justice,* Gabriola Island, BC, Canada: New Society.

Ladson-Billings, G. 2006. Yes, but how do we do it? Practicing culturally relevant pedagogy. In *White Teachers/Diverse Classrooms: A guide to building inclusive schools, promoting high expectations, and eliminating racism,* eds. J. Landsman, and C. W. Lewis, 29–42. Sterling, VA: Stylus.

Landesman, J. and C. Lewis, eds. 2006. *White teachers/diverse classrooms.* Sterling, VA: Stylus.

APPENDIX C

Lederman, N. G. 1992. Students' and teachers' conceptions of the nature of science: A review of the research. *Journal of Research in Science Teaching* 29: 331–359.

Lewis, C. E. 2002. *Lesson study: A handbook of teacher-led change.* Philadelphia: Research for Better Schools.

Libarkin, J. C., and J. P. Kurdziel. 2003. Research methodologies in science education: Gender and the geosciences. *Journal of Geoscience Education* 51 (4): 446–452.

Licht, B.G., and C. S. Dweck. 1984. Sex differences in achievement orientations: Consequences for academic choices and attainments. In *Sex differentiation and schooling,* ed. M. Marland. London: Heinemann.

Linn, M. C., and J. S. Hyde. 1989. Gender, mathematics, and science. *Educational Researcher* 18: 17–27.

Loucks-Horsley, S., P. Hewson, N. Love, and K. Stiles. 2003. *Designing professional development for teachers of science and mathematics.* 2nd ed. Thousand Oaks, CA: Corwin.

Martinez, M. E. 1992. Interest enhancements to science experiments: Interactions with student gender. *Journal of Research in Science Teaching* 29: 167–177.

McIntosh, P. 1990. White privilege: Unpacking the invisible knapsack. *Independent School*, Winter Issue. Excerpted from Working Paper 189. White privilege and male privilege: A personal account of coming to see correspondences through work in women's studies. 1988. Wellesley College Center for Research on women, Wellesley, MA.

Mead, M., and R. Metraux. 1957. Image of the scientist among high school students: A pilot study. *Science* 126: 386–87.

Mueller, C. M., and C. S. Dweck. 1998. Intelligence praise can undermine motivation and performance. *Journal of Personality and Social Psychology* 75, 33–52.

Mullis, I. V. S., M. O. Martin, E. J. Gonzalez, and S. J. Chrostowski. 2004. Findings from IEA's Trends in International Mathematics and Science Study at the Fourth and Eighth Grades, Chestnut Hill, MA: Boston College Press.

National Center for Education Statistics (NCES). 2002. *Digest of education statistics 2001.* Washington, DC: U.S. Department of Education, Office of Educational Research and Improvement.

Literature Cited

National Research Council (NRC). 1996. *National Science Education Standards.* Washington, DC: National Academy Press.

National Research Council (NRC). 2000. *Inquiry and the National Science Education Standards.* Washington, DC: National Academy Press.

National Science Foundation (NSF). 2007. *Women, minorities, and persons with disabilities in science and engineering: 2007,* NSF 07-315. Arlington VA: National Science Foundation, Division of Science Resources Statistics.

Nielsen, L., and M. Long. 1981. Why adolescents can't read: Locus of control, gender, and reading abilities. *Reading Improvement* 18 (4): 339–45.

Oakes, J. 2005. *Keeping track: How schools structure inequality,* 2nd ed. New Haven, CT: Yale University Press.

Peterson C., S. Maier, and M. Seligman. 1995. *Learned helplessness: A theory for the age of personal control.* New York: Oxford University Press.

Pollock, M. 2004. *Colormute: Race talk dilemmas in an American school.* Princeton, NJ: Princeton University Press.

Posner, G. J., K. A. Strike, P. W. Hewson, and W. A. Gertzog. 1982. Accommodation of a scientific conception: Towards a theory of conceptual change. *Science Education* 66 (2): 211–227.

Resnick, L. 1995. From aptitude to effort: A new foundation for our schools. Reprinted by permission of *Daedalus, Journal of the American Academy of Arts and Sciences: American Education, Still Separate, Still Unequal* 124 (4).

Rosenthal, R., and L. Jacobson. 1992. *Pygmalion in the classroom.* Exp. ed. New York: Irvington.

Rowe, M. B. 1974. Wait time and rewards as instructional variables, their influence in language, logic and fate control. Part 1: Wait time. *Journal of Research in Science Teaching* 11: 81–94.

Rowe, M. B. 1987. Wait time: Slowing down may be a way of speeding up. *American Educator* 11: 38–43, 47.

Sadker, M., and D. Sadker. 1994. *Failing at fairness: How our schools cheat girls.* New York: Simon and Schuster.

APPENDIX C

Science Service. 1998. *Westinghouse Science Talent Search science service data base.* Westinghouse Foundation.

Shulman, L. S. 1986. Those who understand: Knowledge growth in teaching. *Educational Researcher* 15 (2): 4–14.

Singleton, G. E., and C. Linton. 2005. *Courageous conversations about Race: A field guide for achieving equity in schools.* Thousand Oaks, CA: Corwin.

Stahl, R .J. 1994. *Using "think-time" and "wait-time" skillfully in the classroom.* ERIC Clearinghouse for Social Studies/Social Science Education, Bloomington, IN. (ERIC Document Reproduction no. ED 370 885). Available online at *http://atozteacherstuff.com/pages/1884.shtml.*

Steele, C. M. 1997. A threat in the air: How stereotypes shape intellectual identity and Performance. *American Psychologist* 52: 613–629.

Steele, C. M., and J. Aronson. 1995. Stereotype threat and the intellectual test performance of African-Americans. *Journal of Personality and Social Psychology* 69: 797–811.

Steinke, J. 1999. Women scientist role models on screen; A case study of contact. *Science Communication* 21 (2): 111–136.

Tanner, K. D., L. Chatman, and D. Allen. 2003. Approaches to biology teaching and learning: Science teaching and learning across the school-university divide—cultivating conversations through scientist-teacher partnerships. *Cell Biology Education* (2): Winter. At *www.lifescied.org/cgi/content/full/2/4/195.*

Tobin, K. and P. Garnett. 1987. Gender related differences in science activities. *Science Education* (71): 91–103.

Vetter, B. 1992. Ferment: Yes—Progress: Maybe—Change: Slow. *Mosaic* 23: 34–40.

APPENDIX D

Author Biographies

University of California at San Francisco (UCSF) Science and Health Education Partnership (SEP) Triad Team

Elizabeth (Liesl) S. Chatman, Science Museum of Minnesota
Liesl Chatman has served in significant leadership positions in university, school district, and museum settings. Currently, she is the Director of Professional Development at the Science Museum of Minnesota. Prior to her work with the Museum, she oversaw science for the Saint Paul Public Schools and served as Executive Director of SEP from 1993 to 2002 where she was the Principal Investigator for the Triad efforts. In addition, she has served as PI or Co-PI on major awards addressing professional development, partnership, and diversity issues from the National Science Foundation, the National Institutes of Health, the Howard Hughes Medical Institute, the California Science Project, 3M, and the Medtronic Foundation. Liesl is also an inveterate graphic journaler and has been visually chronicling science education reform since the early 1990s.

Katherine Nielsen, University of California at San Francisco
Katherine Nielsen is Co-Director of the UCSF Science & Health Education Partnership (SEP) and is the Principal Investigator on numerous grants, including ones from the National Institutes of Health and the Howard Hughes Medical Institute. She has developed and coordinated a variety of partnership programs, including one that integrates English language learning with science. Katherine was a UCSF SEP staff member from 1995–1997 and returned again in 2001. In the intervening years, she taught science at a rural middle and high school and at a tribal college in Montana. During her initial tenure at SEP, Katherine was responsible for implementing the 1994 original Triad design and expanding the number of participating schools. She is currently organizing a collaborative effort to draft a book on science education partnerships.

APPENDIX D

Erin J. Strauss, Science Museum of Minnesota

Erin Strauss is a Professional Development Project Lead at the Science Museum of Minnesota. Prior to this position, she was a Science Instructor at the Perpich Center for Arts Education where she taught Chemistry and Science in Society to 11th and 12th grade arts students. Prior to her position at the Perpich Center, Erin was a Senior Academic Coordinator at SEP. She held a lead role within Triad, particularly with respect to the design of the professional-development series. In addition, she was a Co-Site Director for the California Science Project, was instrumental in the design of an HHMI-funded effort, and pioneered action research and Reflective Teaching Groups at SEP. Prior to coming to SEP in 1997, Erin coordinated science education programs at the Museum of Life and Science in North Carolina and worked as an elementary science coordinator in the Durham County Public Schools.

Kimberly D. Tanner, San Francisco State University

Kimberly Tanner is an Assistant Professor of Biology at San Francisco State University and is the founder of the Science Education Partnership and Assessment Laboratory, which studies science education partnerships and biology conceptual development. From 1997–1998 and 2000–2003, Kimberly served as an SEP Senior Academic Coordinator. At SEP, she was a Co-Site Director for the California Science Project and was instrumental in the design and funding of Triad. From 1998–2000, she studied the influence of scientist-teacher partnerships on scientists through an NSF Postdoctoral Fellowship, studies that continue in her laboratory today. Kimberly is currently Principal Investigator on an NSF GK–12 Partnership Program Award and an NIH Science Education Partnership Award. She is a founding member of the Editorial Board for *Cell Biology Education: A Journal of Life Sciences Education,* sits on the advisory board of the Expanding Your Horizons Network, and regularly serves on committees and review panels for the National Science Foundation, the National Research Council, the Society for Neuroscience, and the American Society for Cell Biology.

Stanford University Evaluation Team

J Myron Atkin, Stanford University

Professor Emeritus J Myron (Mike) Atkin of the Stanford University School of Education led the Triad evaluation efforts and has continued to work with UCSF SEP as an evaluator. He taught science for seven years in New York elementary and secondary schools, then joined the faculty of the University of Illinois at Urbana-Champaign.

After 24 years at Illinois, he moved to Stanford. In both places, he served as dean of education. He is a National Associate of the National Academy of Sciences, where he was a member of the committee that developed the National Science Education Standards and Chair of the Committee on Science Education K–12.

Marjorie Bullitt Bequette, Science Museum of Minnesota

Marjorie Bullitt Bequette is a Professional Development Project Lead at the Science Museum of Minnesota. Prior to this position, she was a post-doctoral research fellow and lecturer in Educational Psychology at the University of Minnesota where she worked on several evaluations of local and national science teacher recruitment and retention projects. As a graduate student in the Stanford University School of Education, she was a research assistant on the evaluation of Triad for three years. Her work on the Triad evaluation and in her dissertation focused particularly on the experiences of students who work directly with scientists through partnership efforts. She is a former middle and high school science teacher.

Michelle Phillips, Inverness Research

Michelle Phillips works for Inverness Research as an evaluator. As a graduate student in the Stanford University School of Education, Michelle played a key role in the evaluation of Triad. In addition, Michelle's thesis explored the impact of partnership programs, including Triad, on scientist volunteers. Her thesis is titled "Their Learning Lab: Novice Scientists Reconciling the Self and the Subject through a Scientist-Teacher Partnership Program." Currently, Michelle is an education researcher and evaluator whose primary interests involve partnerships and informal education initiatives aimed at increasing the public's understanding of and interest in science and mathematics, as well as improving science and mathematics education for underrepresented populations. Most recently, she served as the Senior Researcher for the Center for Informal Learning and Schools (CILS), which is a National Science Foundation–funded partnership among the Exploratorium in San Francisco, King's College London, and the University of California at Santa Cruz. Prior to joining CILS, she was a Social Science Researcher at the Center for Technology in Learning at SRI International in Menlo Park, California. Her early work included conducting research as a marine mammalogist and teaching middle school math and science for four years.

Acknowledgments

This book would not have been possible without the dedicated effort of all the Triad teachers, scientists, students, and SEP staff who worked with us over the nine years of the project. The most heartfelt thanks to all of you for your contributions of heart, body, mind, and spirit as we worked together to make science classrooms equitable learning environments for all students and to become the educators we aspire to be. This book is representative of the group's diverse expertise, interests, passions, and gifts. We hope it will serve both as a record of our community's labor and as an open invitation to others to join us in transforming learning environments for all students.

We also wish to thank the National Science Foundation for its generous financial support of Triad; the University of California, San Francisco (UCSF) for its contributions of space, administrative support, and time for the involvement of scientists and scientific trainees in K–12 science education; and finally the San Francisco Unified School District for its sustained commitment to excellence in K–12 science education for all students and eagerness to partner with the scientific community through the Science and Health Education Partnership (SEP) at the UCSF. In addition to the authors, many past and present SEP staff and Stanford graduate students have contributed to the ideas in this book, including: Patricia Caldera, Helen Doyle, Veena Kaul, Steve Ribisi, Tracy Stevens, Elisa Stone, and Lisa Weasel.

Triad Schools

21st Century Academy K-8 School

Alvarado Elementary School

A.P. Giannini Middle School

Aptos Middle School

Benjamin Franklin Middle School

Everett Middle School

Francisco Middle School

Gloria R. Davis Middle School

Herbert Hoover Middle School

Horace Mann Middle School

James Denman Middle School

James Lick Middle School

Lawton Alternative School

Luther Burbank Middle School

ACKNOWLEDGMENTS

Marina Middle School

Martin Luther King Jr. Middle School

Presidio Middle School

San Francisco Community Alternative School

Treasure Island Elementary School

Visitacion Valley Middle School

All schools are in the San Francisco Unified School District.

Triad Teachers

Adam Singer
Alexandra Gonzales
Ann Dee Clemenza
Barbara Mathews
Beth Simmons
Bo DeAvila
Bonnie Daley
Carol Cockburn
Carol Fields
Cathy Christensen
Cathy Perez
Chanmony Prak
Claudia Scharff
Dan Lazar
David Brody
Debbie Farkas
Deborah Faigenbaum

Elisa Poulos
Elizabeth Abrahams
Emily Lewis
Emma Jones
Eric Hendy
Erla Hackett
Gladys Dalmau
Gloria Andres
Gustavia Gash
Irene Hirota
Jane Gerughty
Jay Cunningham
Jennifer Bahm
Jennifer Porter
Judy Logan
Julie Habeeb
Julie Zastrow

Karen Clayman
Karen Heil
Karen Polk
Kelly Quinn
Kim Coates
Kristen Sorensen
Kyle Isacksen
Linda Payne
Lisbeth Benninger
Lorraine Perry
Louise Cawthon
Madelyn Van Meerbeek
Marge Hazelton
Marlies Lewis
Matt Chapman
Michael Fox

Mie-ling Wiedmeyer
Mishwa Lee
Nathan Draper
Patricia Kudritzki
Patty Golumb
Poe Asher
Raul Amador
Rebecca Pollack
Regan Brooks
Richard Delwiche
Robin Sharp
Sally Meneely
Sandy Jackson
Susan Floore
Tareyton Russ

Acknowledgments

Triad Scientists

Adriana Rossi
Andy Walsh
Angela DePace
Anne Churchland
Bryony Wiseman
Cari Whyne
Cathy Garabedian
Christa Nunes
Christa Tobey
Christelle Sabatier
Christina Cuomo
Christine Rozanas
Christie Fanton
Cynthia Fowler
Deda Gillespie
Delia Garigan
Devi Padmanaban
Devin Parry
Donna Ebenstein
Dung Nguyen
Elena Levine
Ellen Kuwana
Ellie Heckscher
Emily Troemel
Emily Walsh
Erin Gensch

Erin Lopes
Erin Peckol
Essia Buzamondo
Fay Shamanski
Gretchen Ehrenkaufer
Henry Haeberle
Hien Le
Holly Field
Ingrid Ghattas
Irene Yun
Janet Iwasa
Janine Morales
Jennifer Dockter
Jennifer Pickering
Jennifer Turner
Joanne Penko
Julia Kantor
Julia Owens
Julie Blake
Julie Johnson
Kam Dahlquist
Karen Chew
Karen Kirk
Karen Oegama
Katja Brose
Katy Korsmeyer

Kimberly Tanner
Laura Romberg
Leslie Kenna
Linda Yuschenkoff
Lindsay Hinck
Lisa Clement-Ferrill
Lisa Goodrich
Lisa Kim-Shapiro
Maki Inada
Mallika Singh
Maria Gallegos
Maria Wilson
Mehdi Nosrati
Michelle Korenstein
Mika Godzich
Mona Abdallah
Monica Wilhelm
Monique Hultner
Nicole Rank
Noelle Dwyer
Pam Blumson
Patricia Caldera
Patricia Tsao
Polly Shrewsbury
Rachel Brem
Rajneesh Nath

Ranyee Chiang
Reba Howard
Richard Shanks
Robin LeWinter
Sandra Canchola
Sarah Mutka
Savannah Partridge
Sharon Stranford
Sheila Jaswal
Sheri Dorsam
Sherry LaPort
Simon Chan
Steve Ribisi
Sumita Chowdhury-Ghosh
Susan Kirch
Susan Younger
Susanna Mlynarczyk-Evans
Tamara Brenner
Tejal Desai
Teresa Chiaverotti
Tiina Sepp
Victoria Carlton
Yoga Srinivasan
Zemer Gitai

The material presented in this book is based upon work supported by the National Science Foundation under Grant Nos. HRD-9355871, HRD-9813926. Any opinions, findings, and conclusions or recommendations expressed in this material are those of the author(s) and do not necessarily reflect the views of the National Science Foundation.

Index

Note: Page numbers in *italics* refer to figures or tables.

African Americans
    low priority for science, 204
    STEM pursuits, 3, 4
    stereotype threat, 191–192
    tracked into low-level classes, 192
Alberts, Bruce, 4–5
Analyzing data. *See* Data analysis
Anxiety, from confusion, strategies for dealing with, 68–69
Atkin, J Myron "Mike," 8
Atlas of Scientific Literacy, xii, 22, 68, 91

Balloon gas activity, 113
Biology
    cow's eye dissection, 59, 88, 129, 202, 213
    fish dissection, 46
    flies, sorting activity, 194
    honeybee activity, 109
    lamb-heart dissection, 159
    mouse dissection activity, 41–42
    real specimens, 39
    seeds, lesson about, 111
    squeamishness, 39, 41–42, 46, 88
    use of sharp tools, 59
    zebra fish project, 220
Black, Paul, 18
Boys
    anxiety from being confused, 67
    concrete experiences in science, 16, 177
    dominance in classrooms, 3, 16, 177, 180, 181, 229
    self-confidence development opportunities, 16
    tool use expertise, 52, 77
    See also Gender inequity; Students
Bridge building activity, 84, 214
Build a Community of Scientists (goal), 157
    See also Scientists

Case studies. *See* Vignettes
Chatman, Liesl, 5
Chemistry
    balloon gas activity, 113
    in forensic analysis activity, 95–96, 142, 154, 165
    as real science, 123–124
Classroom interaction
    boys' dominance, 3, 16, 177, 180, 181
    dominant girls, 8–9, 167
    groupings, single-sex versus mixed sex, 52
    procedural roles in, 211, 219–220
    round-robin discussion, 182
    stereotyped roles in, 237
    teachers' behavior, 3, 167, 177
    teacher's role in, 210
    with uninterested students, 202
    verbal interaction with peers, 177–178
    working in pairs, 159, 167
    See also Encourage Student Voices (goal); Students; Teachers
Classrooms
    activities in other locales, 115
    coed, girls in, 7, 52, 177
    pro-science culture in, 158
    status in, 209–210
    talking in, 196
    wonder lacking in, 107
Cohen, Elizabeth, 18, 209
Complex instruction, 209
Confidence, fate control and, 80
Confidence to Explore (goal), 37–38
    lack of opportunity to explore, 37–38
    strategies for, 38
    See also Exploration
Confusion
    anxiety and, 67–68
    benefits of, 67
    body language indications, 67
    dealing with anxiety from, 68–69
    as sign of failure, 68
    as sign of learning, 83
    societal attitude toward, 68
    working through problems without teachers' help, 72, 74
    See also Persistence Through Confusion (goal)
Cow's eye dissection, 59, 88, 129, 202, 213
Crime lab activity, 95–96, 142, 154, 165

Danger, in real science, 163
Data analysis, 150
    by Stanford Evaluation Team, 271–272
Data collection
    audiotape, 263
    as 'no fun,' 150
    by Triad staff, 271–272
    videotapes, 9, 263
Debate, in science, 91–92
Defending a Position (goal), 91–92
    motivating girls in, 154
    See also Discourse
Delegate Responsibility (goal), 209–211
    strategies for, 210–211
Designing Professional Development for Teachers of Science and Mathematics, 14, 243
Discourse
    defending a graph, 97
    in forensic analysis activity, 95–96
    girls' socialization in, 18, 91–92
    round-robin discussion, 182, 248
    science discussion among teachers, 117
    speaking up in class, 177–178, 182, 248
    See also Defending a Position (goal); Questions; Science teaching
Do Science to Learn Science (goal), 121–122
    See also Hands-on activities
Drilling holes activity, 54
Duckworth, Eleanor, 107
Dweck, Carol, 191

Eisner, Elliott, xii–xiii
Electromagnetic lesson, 212
Encourage Student Voices (goal), 177–178
    strategies, 178–179
    See also Classroom interaction
Engineering, failure mode testing in, 7
Equity
    discomfort in talking about, 224
    discussions about, 231, 235
    diverse backgrounds and, 167, 224
    family resources and, 233
    mediating subtly, 227
    in mobiles activity materials, 233, 235, 254
    See also Gender inequity; Make Equity Explicit (goal); Stereotypes
Evidence
    discussion of, 150

INDEX

focusing on, 137
ignored by students, 154
reproducible, 147
in science process, 147
using, students' opportunities for, 148
See also Use Evidence to Predict, Explain, and Model (goal)
Expectations (of student performance). *See* Maintain High Expectations (goal); Stereotypes; Teachers
Experiments, step-by-step instructions and, 44
Experiments. *See* Hands-on activities
Exploration
as success in itself, 38
See also Confidence to Explore; Confidence to Explore (goal)

Facilitation
gender inequity, how to handle, 239–240
guidelines, 267–270
overparticipating, 239–240
Failure
competition and, 7, 84
confusion as sign of, 68
fate control, 79–80
girls' fear of, 63
importance in science, 79
providing opportunities for, 84
scientific discoveries and, 79
technical failure, 79, 81
unexpected results seen as, 79, 80
working through mistakes, 86
See also Resilience to Failure (goal)
Familiarity with Tools (goal), 51–53
encouraging girls in, 52–53
See also Tool usage
Family Science Nights, 187, 202, 216
Fate control, 79–80
Fenema, Elizabeth, 37
Field trips, 115
Fish dissection activity, 46
Forensic analysis activity, 95–96, 142, 154, 165

Gender inequity, 218
in adult Triad group, 239
balancing class participation, 180
in doing science activities, 9, 52
in education, 35
explicitly addressing, 224–225

facilitation and, 239–240
fate control, degree of, 79
girls' awareness of, 235
hands-on science opportunities, 121
male help with tool usage, 61, 63, 77
in science, 3, 223
self-confidence development opportunities, 16
status and, 209–210
student goals and, 16–17
teachers' behavior and, 3
teachers' expectations and, 192, 194
Triad clubs and, 5–6, 9
in using tools, 9, 52
See also Boys; Equity; Girls; Make Equity Explicit (goal); Students
Geology, Making mountains activity, 152
Girl Goals. *See* Student Goals
Girls
allowing tool usage by, 53
anxiety from being confused, 67
becoming scientists, 88
behavior in school, 229
defeatism, male help in tool usage and, 63
deferring to male peers, 121
developing strong girls, 9
discourse socialization, 18, 91–92
dominant, in science classes, 8–9, 167
encouraging class participation, 182, 248
equity concerns, 235
fear of failure, 63
high school courses, choices of, 3–4
internalization of disabling stereotypes, 3
learned helplessness, 37, 91
positive feedback needed, 38
rescuing behavior by adults, 76, 198–199
science wonder and, 107
"scientist in me" exercise, 8
self-perceptions as science people, 158
social status decrease with science success, 157
squeamishness, 39, 41–42, 46
step-by-step instructions wanted by, 44
working in pairs, 159, 167

See also Gender inequity; Students
Goal sets, xiii, 14–15, *15*, *23*
as dynamic translation tool, 21
integrated nature of, 19–20
See also Science Goals; Student Goals; Teaching Goals; *specific goals by name*
Gordon, Judy, 37
Graphics. *See* Iconography; Venn diagrams

Hands-on activities
redoing, 214
step-by-step instructions and, 44
student favorites, 70, 127
working in pairs, 159
wrong results, dealing with, 152
See also Do Science to Learn Science (goal); *specific activities by name*
Helping students, 74, 76, 197, 198–199, 210
rescuing girls, 76, 198–199
responding to students' questions, 38, 137
students' exploration of materials, 212
time constraints and, 212
See also Teachers
Helplessness, learned, 37, 91
High school, course-taking patterns, 3–4
Honeybee activity, 109
Hoopsters, 187
Hyde, Janet, 37

Iconography, xii, 30
See also Venn diagrams
Information
assuming experiment outcomes from external sources, 150
being critical of, 136–137
correct answers, focus on, 136, 138, 142
quantity over depth emphasis, 136
from scientists, perceived validity of, 140
students' focus on remembering, 135–137
See also Think Critically, Logically, and Skeptically (goal)
Inquiry, 10
Inquiry. *See also* Science Teaching
Inquiry and the National Science Education Standards, 10, 147

Index

Intelligence
  nature of, 191
  societal views of, 191

Keeping hands in pockets. *See* Helping students

Lamb-heart dissection, 159
Lava lamps activity, 44, 125
Learned helplessness, 37
Lewis, Catherine, 263
Light, colors of, 117
Links, among vignettes, 30
Linn, Marcia, 37
Lotan, Rachel, 209
Loucks-Horsley, Susan, 14, 243
Lowrey, Larry, 13

Maintain High Expectations (goal), 191–193
  See also Stereotypes
Make Equity Explicit (goal), 223–225
  strategies for, 225
  See also Equity; Gender inequity; Stereotypes
Making mountains activity, 152
Measurement, skill in, 58
Microscopes, 56, 211
Mobiles activity, 233, 235, 254
Mouse dissection activity, 41–42
Mousetrap cars activity, 72
Mystery box activity, 127, 202

National Science Education Standards, 18, 105, 121, 173, 210
Nelson, Pinky, 22

Pendulums, experiments, 97, 98
Persistence Through Confusion (goal), 67–69
  See also Confusion
Professional development, xi–xii, 243
  building a reflective community, 265–266
  by classroom research, 262–263
  learning to ask questions of value, 252
  science clubs and, 264–265
  science discussion among teachers, 117
  Triad Framework and, 259–260
  See also Reflect to Improve Practice (goal); Teachers

Questions
  answering questions with, 200
  cultivating the habit of asking, 137
  learning what to ask, 252
  responding to, 38
  in scientists' teaching style, 196
  See also Discourse

Reason
  in science, 107
  students' unwillingness trust in, 142
Reflect to Improve Practice (goal), 243–245
  strategies for, 244
  through writing, 261
  See also Professional development
Resilience to Failure (goal), 7, 79–81
  strategies for, 80–81
  See also Failure
Resnick, Lauren, 18
Roller coaster building activity, 70, 82
Rowe, Mary Budd, 18, 79

Sadker, David, 37
Sadker, Myra, 37
Safety, 86, 163
  sharp tool usage, 59
Science, 209
  chemistry as sole branch of, 123–124
  communication in, 91–92
  danger in, 163
  gender inequity prevalence, 3, 218, 223
  instruments, 51
  predictions in, 147
  "real" science, 39, 123–124, 129, 163–164
  reason in, 107
  as social endeavor, 157, 209
  as solitary endeavor, 157
  student perceptions of, 157, 161, 163–164
  students' lack of interest in, 202
  wonder in, 107–108
Science classes. *See* Classrooms
Science clubs, 5
  African American participation in, 204
  discussing equity in, 231
  driven by dominant girls, 8–9, 167
  as equity teaching laboratories, 7–8
  gender inequity and, 5–6, 9, 93
  girls' ease of learning in, 93
  girls-only clubs, 229
  professional development and, 264–265
  research-based goals, 6
Science fairs, family resources and, 233
Science Goals, xiii, 10–11, 18–19, *19*, *23*, *105*
  strategies, *20*, *26*, *106*, *120*, *134*, *146*, *156*
  See also Goal sets; Triad Framework for Equitable Science Teaching; *specific goals by name*
Science information. *See* Information
Science process, 131, 144, 209
  critical thinking and, 144
  evidence in, 147
  iteration in, 243
Science teaching, *13*, 243
  applying Triad strategies, 246
  of content information, 131
  educational system constraints, 158
  encouraging curiosity, 113
  engaging students in science, 109
  experiments' wrong results, dealing with, 152
  hands-on investigations, 121–122
  not doing things for students, 174
  open-ended investigations, 211
  questioning received information, strategies for, 136–137
  of risk-taking, 81
  strategy exemplars, 20–21, *20*
  student awareness of lesson goals, 125
  Think-Pair-Share strategy, 262
  See also Discourse; Inquiry
Science, Technology, Engineering, and Mathematics (STEM). *See* STEM pursuits
Scientific investigations. *See* Hands-on activities
"Scientist in me" exercise, 8
Scientists
  communities of, 157
  learning how they work, 122
  stereotypes of, 140, 157, 191, 223
  types of questions asked by, 196
  what they do, 161, 163–164, 223
  why girls become, 88
  See also Build a Community of Scientists (goal)

INDEX

Squeamishness, 39, 41–42
Stanford Evaluation Team, 271–272
Status, types of, 210
STEM pursuits
  African Americans' avoidance of, 3, 4
  girls steered away from, 3
Stereotype threat, 191–192
Stereotypes
  African Americans and, 191–192
  Asian students and, 206
  in classroom roles, 237
  of girls, 229
  of scientists, 140, 157, 191, 223
  teachers' expectations of students and, 192
  two-way nature of, 206
  See also Equity; Maintain High Expectations (goal); Make Equity Explicit (goal)
Stopwatches, proficiency in using, 57
Student Goals, xiii, 6, 16–17, *16*, *23*, *35*
  strategies, *24*, *32*, *36*, *50*, *66*, *78*, *90*
  See also Goal sets; Triad Framework for Equitable Science Teaching; *specific goals by name*
Students
  assuming experiment outcomes from external sources, 150
  awareness of lesson goals, 125
  delegating authority to, 209–211
  determining why activities are liked, 109
  distrust in own logic, 142
  with diverse backgrounds, 167, 224
  favorite activities, 70, 127
  focus on correct answers, 136, 142
  frustrated, dealing with, 82
  helping one another, 72
  interest in "real thing," 129
  perceptions of science, 157–158
  perceptions of scientists, 153–154, 161
  teaching one another, 165, 187, 216
  traits for science success, 35

using evidence, strategies for, 148–149
  See also Classroom interaction
Teachers
  admitting mistakes, 86
  behaviors to boys versus girls, 3, 177
  body language of, 165
  classroom behavior, 3, 167, 177
  encouraging students to help each other, 72, 74
  expectations of students, stereotypes and, 192
  familiarity with tools needed by, 52–53
  feeling undervalued, 99
  letting girls use tools, 53
  mean, 88
  personal insecurities, 47
  role in group interaction, 210
  as sounding board for science problems, 74
  See also Classroom interaction; Helping students; Professional development
Teaching Goals, xiii, 9–10, 17–18, *17*, *23*, 173–174, *173*
  strategies, *25*, *170*, *176*, *190*, *208*, *222*, *242*
  See also Goal sets; Triad Framework for Equitable Science Teaching; *specific goals by name*
Think Critically, Logically, and Skeptically (goal), 135–137
  See also Information
Tool usage
  bulb changing on spectrophotometer, 63
  drilling holes activity, 54
  gender inequity in, 9, 52, 237
  gender socialization and, 51–52, 63
  girls' pride in, 54, 55
  importance of, 51
  male insistence on helping with, 61, 63, 77

  sharp tools, 59
  teachers' need for ease with, 52–53
  woodworking, 61–62, 198–199
  See also Familiarity with Tools (goal)
Tools
  microscopes, 56
  stopwatches as, 57
Triad clubs. *See* Science clubs
Triad community of practice, xi, 4–5
Triad Framework for Equitable Science Teaching, xi, 21
  equity guidelines applied in training for, 188
  evolution of, 13–15, *13*
  gender inequity in adult retreat, 239
  girls' views of, 184–185
  iconography, xii, 21, 30
  methodology, 271–272
  strategy exemplars, 20–21, *20*
  See also Goal sets; Science Goals; Student Goals; Teaching Goals

University of California at San Francisco (USCF), Science and Health Partnership (SEP), 4–5
Use Evidence to Predict, Explain, and Model (goal), 147–149
  See also Evidence

Venn diagrams, xii, 30
  See also Iconography
Vignettes, xi, 29
  links, 30

Wonder
  as a skill, 108
  correct answers more important than, 136
  encouraging student curiosity, 113
  lacking in science classes, 107
  in science, 107–108
Woodworking activities, 61–62

Zebra fish project, 220